Digital Beamforming in Wireless Communications

For a complete listing of the *Artech House Mobile Communications Library*,
turn to the back of this book.

Digital Beamforming in Wireless Communications

John Litva and Titus Kwok-Yeung Lo

Artech House
Boston • London

Library of Congress Cataloguing-in-Publication Data
Litva, J. (John), 1937-
Digital beamforming in wireless communications/John Litva and Titus Kwok-Yeung Lo.
p. cm.
Includes bibliographical references and index.
ISBN 0-89006-712-0 (alk. paper)
1. Wireless communication systems. 2. Electron beams. 3. Digital communications. I. Lo, Titus Kwok-Yeung. II. Title.
TK5103.2.L58 1996
621.3845—dc20 96-27151
CIP

British Library Cataloguing in Publication Data
Litva, J. (John), 1937-
Digital beamforming in wireless communications
1. Digital communications
I. Title II. Lo, Titus Kwok-Yeung
621.3'82

ISBN 0-89006-712-0

To the countless scientists and engineers for their original contributions to digital beamforming development.

Cover design by Kara Munroe-Brown

685 Canton Street
Norwood, MA 02062

International Standard Book Number: 0-89006-712-0
Library of Congress Catalog Card Number: 96-27151

10 9 8 7 6 5 4 3

Contents

Preface

Spatial processing - the last frontier. . .

Digital beamforming (DBF) is a technology that has been both spawned and incubated by the sonar and radar communities. It now appears that its ultimate technological home will be in wireless communications, more specifically, personal communications services, local multipoint distribution systems, and satellite. This then is an example of a technology, which although initially was developed for a number of remote sensing applications, will ultimately find its true expression in a totally different area, one more closely aligned with the development of the information highway.

A DBF antenna can be considered the ultimate antenna in the sense that it can capture all of the information that falls on the antenna's face, and can apply whatever signal processing is required to make the information of interest to the observer, presented in such a way as to make it most readily understood. Although a digital beamforming antenna is the ultimate antenna, for the most part DBF has taken a back seat to other antenna technologies in so far as practical applications are concerned. The application of DBF has been largely waiting for the speed of microprocessors to reach the speeds of present-day digital signal processing chips. For the most part, even some twenty years ago, workers understood what the processing requirements were to make real-time beamforming a reality. During this incubation period, based on the current state of microprocessors, and using Moore's law, which states that the speeds of microprocessors double every 18 months, one could have made a quick calculation of when the capability of these processors would be equal to the task of DBF processing. For example, one could have predicted some time ago that a company would be introducing DBF technology to satellite systems in 1996, as is currently the case with the Inmarsat-P satellites—there is in fact a major aerospace company currently developing DBF technology for introduction into a suite of satellites. Since aerospace companies are by their very nature conservative, this truly signals the coming of age for DBF antenna technology.

DBF technology has in a sense been a precursor of the convergence that is taking place in modern communications, that of communications and computers. More specifically, DBF has been dependent on the convergence of two technology steams, that of radio frequency and software. The DBF antenna can truly be called the product of a marriage between electromagnetics and digital signal processing: electromagnetics is required to develop the communications signal, and the computer software is required to make the communications system smart.

In this book we introduce the important technology of digital beamforming and its upcoming applications in wireless and satellite communications. This book is intended to be a tutorial and reference book for use by communications engineers and technicians who are involved in development and design of wireless communications systems, by managers who are involved in wireless communications system engineering and system planning, by researchers who are involved in wireless communications research, and by those who are working in related fields and interested in DBF technology and its applications. The readers of this book are required to have some basic knowledge in communications. This book consists of three parts. The first part, which includes Chapters 1 to 4, introduces the concepts and principles of DBF. Chapter 1 gives a general introduction to the background of DBF and the rationale for using DBF in wireless communications. Chapter 2 provides the fundamentals of DBF and Chapter 3 presents some basics of adaptive beamforming. Chapter 4 examines the effects of errors on DBF. The second part discusses the applications of DBF to wireless communications and comprises Chapters 5 to 9. Chapter 5 introduces the concept of spatial processing in wireless communications, including space-division multiple access, while Chapter 6 describes different configurations of DBF networks. The applications of DBF to satellite mobile communications, land mobile communications, and indoor and data communications are addressed in Chapters 7, 8, and 9, respectively. The last part of this book, consisting of Chapters 10, 11, and 12, reviews the state-of-the-art of the supportive technologies for DBF, including antenna elements, transceiver technology, and digital-signal-processing technology. It is our hope that this book will help its readers derive the necessary information required to develop an understanding of the technology and to join the exciting quest of using spatial signal processing to improve the performance of communications systems. In terms of performance improvement, time domain processing techniques have virtually been squeezed down to their last one-tenth of a decibel. Spatial domain processing is truly the last frontier in terms of the rewards that can be achieved in improving the performance of wireless communications systems.

We would like to thank Dr. M. A. Bree, QCC Communications Corporation, Saskatoon, Mr. R. Ho, IBM Canada, for their permission to reproduce a number of figures. We are also grateful to the Institute of Electrical and Electronic Engineers

(New York) for giving us permission to reproduce certain figures. The following individuals are deeply appreciated for their invaluable contributions in various ways: Mr. A. Chen of McMaster University, Mr. C. Laperle of Laval University, Ms. A. Sandhu of McMaster University, Dr. Y. Shen of Harris Farinon Canada, Mr. X. Yu of McMaster University, Ms. M. Zhang of McMaster University, and Dr. W. Zang of Xian Institute of Space Radio Technology, China. We are pleased to acknowledge the Telecommunications Research Institute of Ontario (TRIO), Natural Science and Engineering Research Council (NSERC) of Canada, and Defense Research Establishment of Ottawa (DREO) for their support of the research in DBF at the Communications Research Laboratory.

John Litva
Titus K. Y. Lo

August 1996

Chapter 1

Introduction

1.1 WIRELESS COMMUNICATIONS IN DEMAND

The 1990s have been described as the decade of wireless telecommunications. The evolution of telecommunications, from the wired phone to personal communications services (PCS), is resulting in the availability of wireless services, which were not previously considered practical. In providing different types of wireless services, such as fixed, mobile, outdoor, indoor, and satellite communications, the wireless communications industry is experiencing an explosive growth. The total end-user expenditures in North America for wireless services were expected to exceed $17 billion (U.S.) in 1992. According to an estimate by the International Telecommunications Union, a United Nations agency, the world spent $535 billion (U.S.) on telecommunications in 1992. This figure is growing by more than 10% a year.

In mobile communications, the rapid increase in the number of cellular phones demands an ever increasing capacity for future mobile communications systems. At the beginning of 1990, there were 4.5 million cellular telephones, 6.5 million pagers, and 25 million cordless telephones in the United States alone [1]. By the end of 1992, the number of cellular phones in United States had increased to over 11 million, representing a 150% jump in three years [2]. The total production of cordless phones will exceed 200 million annually by the turn of the century, and 300 million by 2010 [3]. Even though the mobile communications industry is still in a growth phase, a new generation of communications service is emerging and fast becoming a force in the mainstream of life—namely PCS. The services provided by future wireless PCS will far exceed today's mobile system because PCS will consist of an extensive system of radio networks carrying both voice and data. Ultimately, PCS will become more sophisticated and widespread than mobile or cellular systems. When mature, PCS may comprise a global wireless network that could identify users anywhere in the world and reach individuals regardless of their locations—whether on board an airplane, in an elevator, or walking a dog on the street. Because PCS will provide pervasive communications services, it will require much higher levels of system capacity than today's mobile or cellular systems.

One of the missions of PCS is to provide service anywhere in the world. Satellites will play a significant and unique role in carrying out this mission. Satellites can offer services to users in rural and remote areas that are outside the range of terrestrial telecommunications systems. Mobile or personal communications services that are to be provided via satellites represent a relatively new field of investigation for satellite communications engineers. Such services are becoming available for aeronautical, maritime, and land mobile applications, whereby communications are being facilitated between terrestrial networks and aircraft, ships, and mobile vehicles. Through the use of digital mobile satellite communications systems, users anywhere in the world will be able to access voice and data communications through global networks while traveling as passengers in airplanes, trains, or ships. At the present time, a number of communications satellite systems are being designed and built to provide wide-ranging communications services. These include INMARSAT and MSAT, which occupy the geosynchronous orbits (GEO); ODYSSEY, which operates at a medium-altitude earth orbit (MEO); and IRIDIUM and GLOBALSTAR, which are designed for low earth orbits (LEO). However, these new services create an increasing demand for higher capacity communications satellites.

1.2 MULTIPLE ACCESS

A higher demand in wireless communications calls for higher systems capacities. The capacity of a communications system can be increased directly by enlarging the bandwidth of the existing communications channels or by allocating new frequencies to the service in question. However, since the electromagnetic spectrum is limited, thereby making it a valuable resource, and the electromagnetic environment is increasingly becoming congested with a proliferation of unintentional and intentional sources of interference, it may not be feasible in the future to increase system capacity by opening new spectrum space for the wireless communications applications. Therefore, efficient use of the frequency resource is critical if communications engineers are to increase the capacity of a communications system.

Over the last four decades, engineers have made considerable progress in increasing the capacity of a wireless communications system, especially for digital wireless communications. We are now able to achieve transmission rates approaching within a factor of 2 of the maximum rate determined by Shannon's capacity formulas. Coupling the benefit of source coding and channel coding, digital techniques can significantly reduce either the required transmitter power or the bandwidth of the transmission, or both, and yet achieve better performance or quality than analog systems can. Furthermore, with digital techniques, frequency efficiency can be improved by using multiple access techniques to provide high system capacity. Multiple access refers to the simultaneous transmission by numerous users to or through a common receiving point. Basically, it means that a number of users, which in this case are represented by randomly dispersed communicators with handsets, have the desire to use or have access to a scarce resource, namely, the basestation. Multiple access is implemented by having the users share the basestations. There are four domains in which sharing can take place. That is, access can be apportioned to one or more of the following domains: (1) bandwidth, (2) time, (3) code, or (4) space. In frequency-division multiple access (FDMA), the frequency spectrum is divided into segments that are apportioned among different users. With the arrival of digital techniques, time-division multiple access (TDMA) became a practical access technology. Here, each user is apportioned the entire transmission resource periodically for a brief period of time. The users' transmissions are therefore intermittent in nature, a condition that can only be accommodated by a digital transmitter that can store its source bits and then send them out at a transmission speed higher than that at which they are generated. Digital techniques also enable another multiple access method, code-division multiple access (CDMA). CDMA employs spread spectrum modulation (i.e., each user's digital waveform is spread over the entire frequency spectrum that is allocated to all users of the network). Each of the transmitted signals is modulated with a unique code that identifies the sender. The intended receiver then uses the appropriate code to detect the signal of his or her choice.

Another form of multiple access is known as space-division multiple access (SDMA). Although this term has not seen much use in the open literature, SDMA has in fact been widely used in wireless communications. This may come as a surprise to most readers. For example, in a cellular telephone network, where a large geographical coverage is desired and a large number of mobile transceivers must be supported, the region is divided into a large number of cells. This allows the same carrier frequency to be reused in different cells. In fact, this is a primitive form of SDMA, in that communication signals that are transmitted at the same carrier frequency in different cells are separated by a spatial distance to reduce the level of cochannel interference. In principle, the larger the amount of cells in a region, the higher the level of frequency reuse and hence the higher capacity that can be achieved. This is one of the reasons why microcells and picocells have been proposed for PCS. However, the criterion used for defining cochannel cells is that the distance between them is sufficiently large that intercell interference is lower than some acceptable limit. For a given basestation transmission power level, this puts a limit on the number of cells in a geographical area. To further increase the capacity, communications engineers are realizing that more advanced forms of SDMA are needed. For example, 120° sectorial beams at different carrier frequencies can be used within a cell and each sectorial beam can be used to serve the same number of users as are served in the case of ordinary cells, as illustrated in Figure 1.1. With careful frequency planning, the capacity can be tripled and the carrier-to-interference ratio (CIR) can also be increased. The ultimate form of SDMA, however, is to use independently steered high-gain beams at the same carrier frequency to provide service to individual users within a cell, as shown in Figure 1.2. That is, communications can be simultaneously carried out between users and the basestation. The latest form of SDMA usually employs adaptive antenna arrays, which have recently drawn considerable interest from the communications community [4–9]. In large measure, adaptive array techniques are dependent on *digital beamforming* (DBF) technology for their practical implementation.

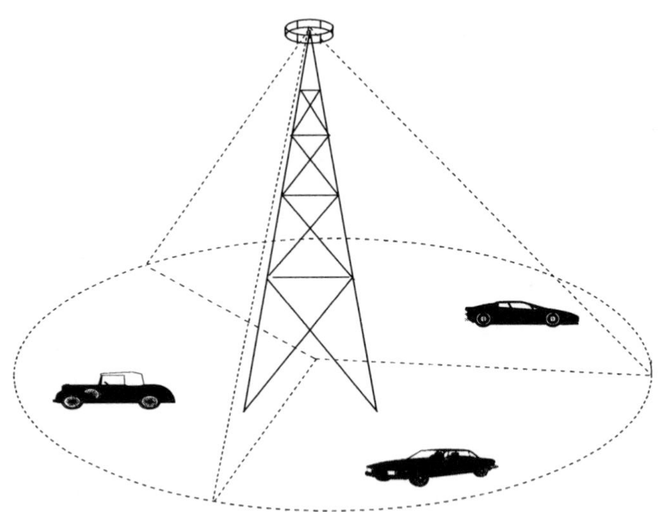

Figure 1.1 120° sectorized cell pattern. A different set of carrier frequencies is used in each of the sectors. These frequencies are reused in other sectors of other cell sites. The frequency reuse pattern is chosen to keep interference to a minimum.

In satellite communications, frequency efficiency can be improved by the use of antenna spot beams. The use of spot beams is indeed a form of SDMA, which allows a communications satellite to simultaneously provide coverage to different regions

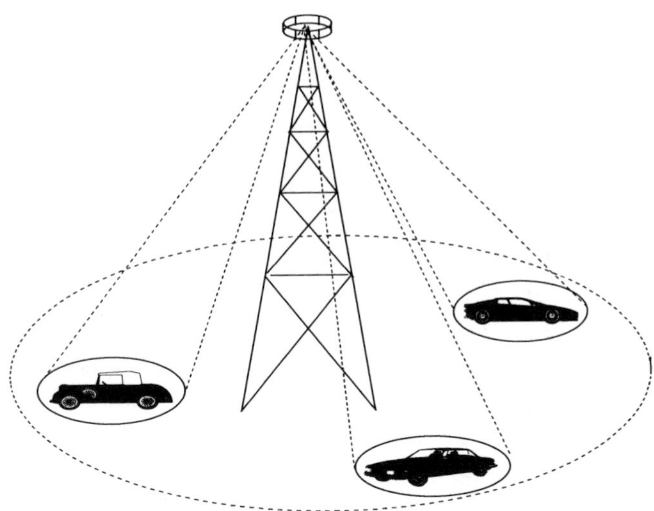

Figure 1.2 Communication is simultaneously carried out between users and the basestation by using independently steered high-gain beams at the same carrier frequency to provide service to individual users within a cell.

using common frequency channels (Figure 1.3); that is, the satellite is reusing frequencies to increase its overall capacity. This approach gives a frequency reuse factor that is normally equal to the number of independent spot beams. Further improvements can be realized by switching the antenna interconnections synchronously with the TDMA frame rate, known as satellite-switched TDMA (SS/TDMA). To maximize the SDMA capacity or the level of frequency reuse, it is desirable to generate a large number of independently steered high-gain mobile spot beams. However, this is difficult to achieve using the analog beamforming technology that is currently used in satellite communications antenna systems. Conventional satellite antennas can only generate a very small number of spot beams with fixed locations and coverage. Thus, the SDMA capacity is very limited. Analog beamforming techniques are subject to an upper limit in terms of their capacity for generating independent high-gain beams. At the present time it is difficult to define the upper limit with any degree of certainty. From a conceptual point of view, we may consider the noise figure for a direct radiating array, with a transceiver located at each element of the array, to be largely determined by the noise figure of the transceiver. If one were to carry out low-level beamforming using multilayer printed circuit technology, theoretically the number of analog beams that could be formed is large. At some point, though, losses in the beamforming network will start to degrade the performance of the beamforming network. It has been reported that by using multilayer stripline technology, 16 controlled beams have been generated based on a multibeam Butler matrix beamforming assembly [10]. It is thought that this technology could be extended to the point where a system might generate as many as 50 beams. Based on intuition, one could suggest that 100 might be an upper limit for an analog beamformer. It seems reasonably safe to conclude that if a satellite system were required to generate 1000 beams, the only practical way to achieve this would be by using a digital beamformer. In recent years, the concept of using DBF technology for applications to future communication satellite payloads [11–16] has been slowly taking root in the technical community. It is envisaged that these satellite systems would support up to 7000 channels. With a frequency reuse factor of 5, such systems would have to generate 1400 independent frequency channels. If each channel were to occupy one beam, up to 7000 independent beams would have to be generated. Digital beamforming is the only technology that would support the generation of that many beams.

Figure 1.3 A communications satellite provides simultaneous coverage to different geographical regions, all at the same carrier frequency, through the use of independent spot beams (shaded circles).

1.3 DIGITAL BEAMFORMING

The early concepts underlying digital beamforming were first developed for applications in sonar [17] and radar systems [18]. DBF represents a quantum step in antenna performance and complexity. It is based on well-established theoretical concepts which are now becoming practically exploitable, largely as a result of recent major advances in areas such as *monolithic microwave integrated circuit* (MMIC) technology and *digital signal processing* (DSP) technology. DBF technology has reached a sufficient level of maturity that it can be applied to communications for improving system performance. The application of DBF to wireless communications is no longer only a theoretical possibility. It is fast becoming a reality. Furthermore, the demand for increased capacity is a major driving force for incorporating DBF into future wireless communications systems.

DBF is a marriage between antenna technology and digital technology. A generic DBF antenna system shown in Figure 1.4 consists of three major components: the antenna array, the digital transceivers, and the digital signal processor.

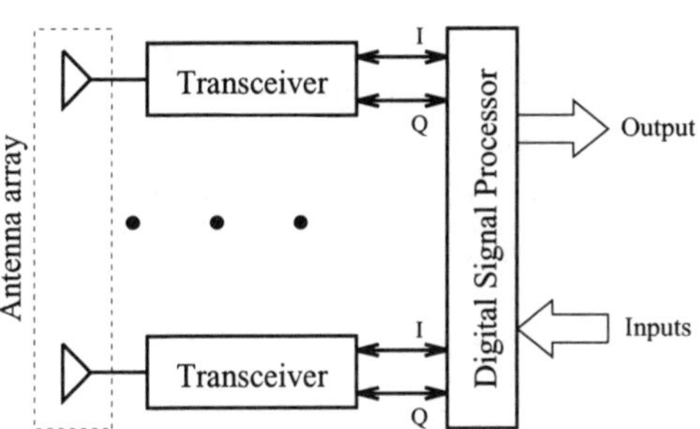

Figure 1.4 A generic DBF antenna system.

In a DBF antenna system, the received signals are detected and digitized at the element level. By capturing the RF information in the form of a digital stream, we open the door to a large domain of signal processing techniques and algorithms that can be used to extract information from the spatial domain data. Digital beamforming is based on capturing the radio frequency (RF) signals at each of the antenna elements and converting them into two streams of binary baseband signals (i.e., in-phase (I) and quadrature-phase (Q) channels). Included within the digital baseband signals are the amplitudes and phases of signals received at each element of the array. The beamforming is carried out by weighting these digital signals, thereby adjusting their amplitudes and phases such that when added together they form the desired beam. This process can be carried out using a special-purpose digital signal processor. That is, the function, which is usually carried out using an analog beamforming network, is now carried out using a digital processor. This approach preserves the total information available at the aperture, in contrast to an analog beamformer, which produces only the weighted sum of these signals and thus reduces the signal dimensionality from K to 1 (Figure 1.5). The key to this technology is the accurate translation of the analog signals into the digital regime. This is accomplished using complete heterodyne receivers, which must all be closely matched in amplitude and phase. The matching need not be carried out by making adjustments to the hardware, but rather by applying a calibration process in which the values of the data stream are adjusted prior to beamforming. The receivers perform the following functions: frequency down-conversion, filtering, and amplification so that signal levels are commensurate with the input requirements of analog-to-digital converters (ADC). The main advantage to be gained from digital beamforming is greatly added flexibility without any attendant degradation in signal-to-noise ratio (SNR). In many ways it can be considered to be the ultimate antenna, in that all of the information arriving at the antenna face is captured in the digital streams that flow from this face. All of this information is available for processing in the beamformer. There are many possible equipment configurations that can be used to perform the digital processing. Since the beamforming instructions are driven by software routines, there is wide-ranging flexibility in the types of beams that can be produced, including scanned beams, multiple beams, shaped beams, and beams with steered nulls.

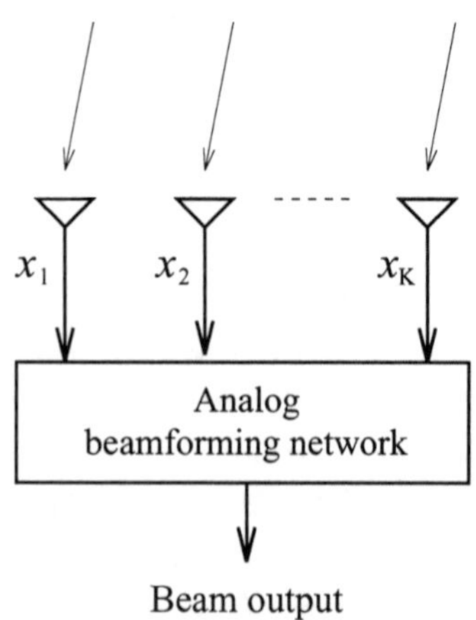

Figure 1.5 An analog beamformer reduces the signal dimensionality from K to 1.

When we combine the access that digital processing has to a higher level of information with its flexibility, we find that digital beamforming allows for a number of attractive features beyond the capabilities of conventional phased arrays:

1. A large number of independently steered high-gain beams can be formed without any resulting degradation in signal-to-noise ratio.
2. All of the information arriving at the antenna array is accessible to the signal processors so that system performance can be optimized.
3. Beams can be assigned to individual users, thereby assuring that all links operate with maximum gain.
4. Adaptive beamforming can be easily implemented to improve the system capacity by suppressing cochannel interference. Any algorithm that can be expressed in mathematical form can be implemented. As a byproduct, adaptive beamforming can be used to enhance the system immunity to multipath fading.
5. DBF systems are capable of carrying out antenna system real-time calibration in the digital domain. Therefore, one can relax the requirements for a close match of amplitude and phase between transceivers, because variation in these parameters can be corrected in real time.
6. Digital beamforming has the potential for providing a major advantage when used in satellite communications. If, after the launch of the satellite, it is found that the performance of the beamformer needs to be upgraded, a new suite of software can be telemetered up to the satellite. This means that the life of the satellite can be expanded by retrofits at various intervals, during which the satellite's capabilities are upgraded.

Adaptive beamforming technology is also referred to as smart antenna technology in some literature. The use of the term "smart" reflects the antenna's ability to adapt to the environment in which it operates. Although "smart antennas" and "adaptive antennas" are interchangeable, throughout the book we will keep the tradition by using "adaptive antennas" or "adaptive beamforming" for a number of reasons. We believe that the term adaptive describes the antenna's ability more concisely and precisely. Furthermore, the term adaptive also reflects the historical development of this particular branch in the antenna field. Finally, the term adaptive beamforming has been also used in many other areas, such as satellite communications, radar, and remote sensing.

References

[1] P. MacLaren and I. Nakonecznyi, "Wireless customer premise equipment," in *Wireless'91 Proc.*, Calgary, Alberta, 1991.

[2] Cellular Telephone Industries Association, *Annual Conf. Proc.*, Dallas, March, 1993.

[3] L. Wirbel, "New markets in a wireless world," *Electronic Engineering Times*, vol. 734, Feb. 1993.

[4] J. H. Winters, "Optimum combining in digital mobile radio with co-channel interference," *IEEE Trans. Veh. Technol.*, vol. 33, pp. 144–155, Aug. 1984.

[5] S. C. Swales, M. A. Beach, D. J. Edwards, and J. P. McGeehan, "The performance enhancement of multibeam adaptive basestation antennas for cellular land mobile radio systems," *IEEE Trans. Veh. Technol.*, vol. 39, pp. 56–57, Feb. 1990.

[6] S. Anderson, M. Millnert, M. Viberg, and B. Wahlberg, "An adaptive array for mobile communications systems," *IEEE Trans. Veh. Technol.*, vol. 40, pp. 230–236, Feb. 1991.

[7] W. C. Y. Lee, "Applying the intelligent cell concept to PCS," *IEEE Trans. Veh. Technol.*, vol. 43, pp. 672–679, Aug. 1994.

[8] A. F. Naguib, A. Paulraj, and T. Kailath, "Capacity improvement with base-station antenna arrays in cellular CDMA," *IEEE Trans. Veh. Technol.*, vol. 43, pp. 691–698, Aug. 1994.

[9] M. Barrett and R. Arnott, "Adaptive antennas for mobile communications," *Electron. Comm. Eng. J.*, vol. 6, pp. 203–214, Aug. 1994.

[10] B. Worden, "Anaren beamforming technology: An anaren white paper," Technical report, Anaren Microwave, Inc., Feb. 1994.

[11] A. D. Craig et al., "Study on digital beamforming networks," Technical Report TP 8721, British Aerospace Ltd., Stevenage, England, July 1990.

[12] M. Barrett, "Digital beamforming network technologies for satellite communications," in *ESA Workshop on Advanced Beamforming Networks for Space Applications*, Noordjwijk, Netherlands, 1991.

[13] F. J. Lake and R. P. Curnow, "Active interference suppression and onboard source location within a digital beamforming payload," in *ESA Workshop on Advanced Beamforming Networks for Space Applications*, Noordjwijk, Netherlands, 1991.

[14] A. D. Craig et al., "A digital beamforming payload concept for advanced mobile missions," in *ESA Workshop on Advanced Beamforming Networks for Space Applications*, Noordjwijk, Netherlands, 1991.

[15] A. D. Craig, C. K. Leong, and A. Wishart, "Digital signal processing in communications satellite payload," *Electron. Comm. Eng. J.*, vol. 4, June 1992.

[16] H. G. Gockler and H. Eyssele, "A digital FDM-demultiplexer for beamforming environment," *Space Comm.*, vol. 10, pp. 197–205, 1992.

[17] T. E. Curtis, "Digital beam forming for sonar system," *IEE Proc. Pt. F*, vol. 127, pp. 257–265, Aug. 1980.

[18] P. Barton, "Digital beamforming for radar," *IEE Proc. Pt. F*, vol. 127, pp. 266–277, Aug. 1980.

Chapter 2

Fundamentals of Digital Beamforming

Digital beamforming is based on capturing the RF signals the antenna elements and converting them into two streams of binary baseband I and Q signals, which jointly represent the amplitudes and phases of signals received at the elements of the array. The beamforming is carried out by weighting these digital signals, thereby adjusting their amplitudes and phases such that when added together they form the desired beam. Basically, a digital beamforming system consists of an array of antenna elements, independent receivers for the individual antenna elements, and a digital signal processor. In order for readers to understand the basic principles of digital beamforming, we present its fundamentals which include antenna arrays and beamforming processing.

2.1 INTRODUCTION TO ANTENNA ARRAYS

In many applications of antennas, point-to-point communications is of interest. A highly directive antenna beam can be used to advantage. The direction beam can be realized by forming an array with a number of elemental radiators. As the directivity of the antenna increases, the gain also increases. At the receive end of the communications link, the increase in directivity means that the antenna receives less interference from its signal environment. Also, for the same signal level at the receive antenna, if one increases the gain by a factor of, say, 10, one can reduce the transmitted power tenfold. It is clear that high-gain antennas have significant advantages in point-to-point communications circuits.

2.1.1 Basic Antenna Array Parameters

Terms and definitions commonly used in the study of antennas and arrays can be found in [1–3]. Here we will provide some of the parameters and definitions that are particularly relevant to issues addressed in this book. Although most of the parameters are defined in terms of transmitting antennas, reciprocity ensures that these definitions are also applicable to receiving antennas.

Radiation Pattern

The relative distribution of radiation power as a function of direction in space is called the radiation pattern of an antenna.

Array Factor

The array factor represents the far-field radiation pattern of an array of isotropically radiating elements. The array factor will be denoted, throughout this book, by $F(\phi, \theta)$, where ϕ represents the azimuth angle and θ represents the elevation angle in space.

Main Lobe

The main lobe of an antenna radiation pattern is the lobe containing the direction of maximum radiation power.

Sidelobes

Sidelobes are lobes in any direction other than that of the main lobe. For a linear array with uniform weighting, the first sidelobe (i.e., the one nearest the main lobe) in the radiation pattern is about 13 dB below the peak of the main lobe.

Beamwidth

The beamwidth of an antenna is the angular width of the main lobe in its far-field radiation pattern. Half-power beamwidth (HPBW), or 3-dB beamwidth, is the angular width measured between the points on the main lobe that are 3-dB below the peak of the main lobe. A linear array with uniform weighting has a 3-dB beamwidth of

$$\text{HPBW} = \frac{0.88\lambda}{A} \tag{2.1}$$

where A is the aperture length of the array.

Antenna Efficiency

Antenna efficiency is defined as the ratio of the total power radiated by the antenna and the total power input to the antenna.

Directive Gain

Directive gain is a far-field quantity and defined as the ratio of the radiation density in a particular angular direction in space to the radiation density of the same power radiated isotropically; that is,

$$D(\phi,\theta) = \frac{4\pi \text{ power radiated per unit solid angle in direction } \phi,\theta}{\text{Total power radiated by antenna}} \tag{2.2}$$

Directivity

Directivity is the maximum directivity gain of an antenna; that is, it is the directive gain in the direction of the maximum radiation density.

Antenna Gain

The gain of an antenna is defined as the ratio of the radiation density in a particular angular direction in space to the total input power to the antenna; that is,

$$G(\phi,\theta) = \frac{4\pi \text{ power radiated per unit solid angle in direction } \phi,\theta}{\text{Total input power to antenna}} \tag{2.3}$$

The maximum gain G, or simply gain, is the product of the directivity and the antenna efficiency; that is,

$$G = D\eta \tag{2.4}$$

Effective Isotropic Radiation Power

The effective isotropic radiation power (EIPR) is the product of the input power to the antenna and its maximum gain.

Effective Aperture

The effective aperture of an antenna is defined as the area of an ideal antenna that would absorb the same power from an incident plane wave as the antenna in question. Under matched polarization conditions, the effective aperture is given as

$$A_{\text{eff}} = \frac{\lambda G}{4\pi} \tag{2.5}$$

Aperture Efficiency

The aperture efficiency of an antenna is defined as the ratio of its effective aperture and physical aperture.

Grating Lobe

In an antenna array, if the element spacing is too large, several main lobes will be formed in visible space on each side of the array plane. The extra main lobes formed with large element spacing are referred to as grating lobes.

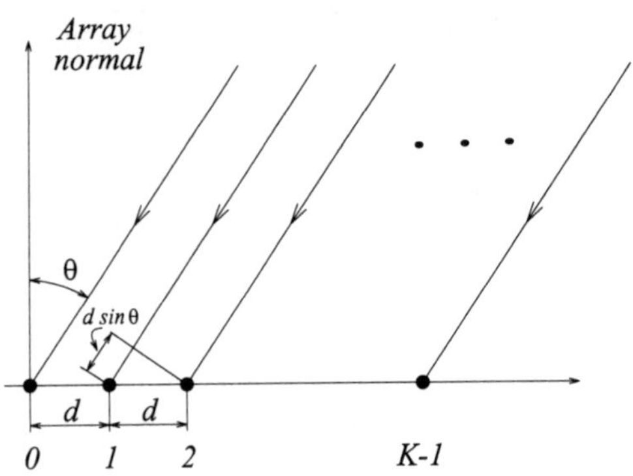

Figure 2.1 A uniformly spaced linear array.

2.1.2 Linear Array

In Figure 2.1 a uniformly spaced linear array is depicted with K identical isotropic elements. Each element is weighted with a complex weight V_k with $k = 0, 1, \cdots, K-1$, and the interelement spacing is denoted by d. If a plane wave impinges upon the array at an angle θ with respect to the array normal, the wavefront arrives at element $k+1$ sooner than at element k, since the differential distance along the two ray paths is $d \sin\theta$. By setting the phase of the signal at the origin arbitrarily to zero, the phase lead of the signal at element k relative to that at element 0 is $\kappa k d \sin\theta$, where $\kappa = \frac{2\pi}{\lambda}$ and $\lambda =$ wave length. Adding all the element outputs together gives what is commonly referred to as array factor F:

$$F(\theta) = V_0 + V_1 e^{j\kappa d \sin\theta} + V_2 e^{j2\kappa d \sin\theta} + \cdots = \sum_{k=0}^{K-1} V_k e^{j\kappa k d \sin\theta} \tag{2.6}$$

which can be expressed in terms of vector inner product:

$$F(\theta) = \boldsymbol{V}^T \boldsymbol{v} \tag{2.7}$$

where

$$\boldsymbol{V} = [V_0\ V_1\ \cdots\ V_{K-1}]^T \tag{2.8}$$

is the weighting vector and

$$\boldsymbol{v} = \begin{bmatrix} 1 & e^{j\kappa d \sin\theta} & \ldots & e^{j(K-1)\kappa d \sin\theta} \end{bmatrix}^T \tag{2.9}$$

is the array propagation vector that contains the information on the angle of arrival of the signal. If the complex weight is

$$V_k = A_k e^{jk\alpha} \tag{2.10}$$

where the phase of the kth element leads that of the $(k-1)^{\text{th}}$ element by α, the array factor becomes

$$F(\theta) = \sum_{k=0}^{K-1} A_k e^{j(\kappa k d \sin\theta + k\alpha)} \tag{2.11}$$

If $\alpha = -\kappa d \sin\theta_0$, a maximum response of $F(\theta)$ will result at the angle θ_0. That is, the antenna beam has been steered towards the wave source. An example of $F(\theta)$ for an eight-element linear array is given in Figure 2.2, where the antenna beam is steered towards the antenna boresight.

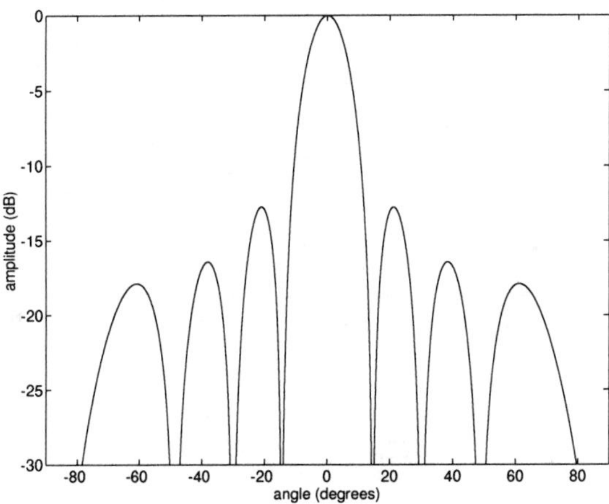

Figure 2.2 The beam pattern of an eight-element linear array.

2.1.3 Circular Array

A circular array consisting of K identical isotropic elements evenly spaced in a circle of radius R is shown in Figure 2.3. Each element is weighted with a complex weight

V_k for $k = 0, 1, \cdots, K-1$. Since the K elements are equally spaced around the circle of radius R, the azimuth angle of the kth element is given as $\phi_k = 2k\pi/K$. If a plane wave impinges upon the array in the direction of (θ, ϕ) in the coordinate system shown in Figure 2.3, the relative phase at the kth element with respect to the center of the array is given by

$$\beta_k = -\kappa R \cos(\phi - \phi_k) \sin\theta \tag{2.12}$$

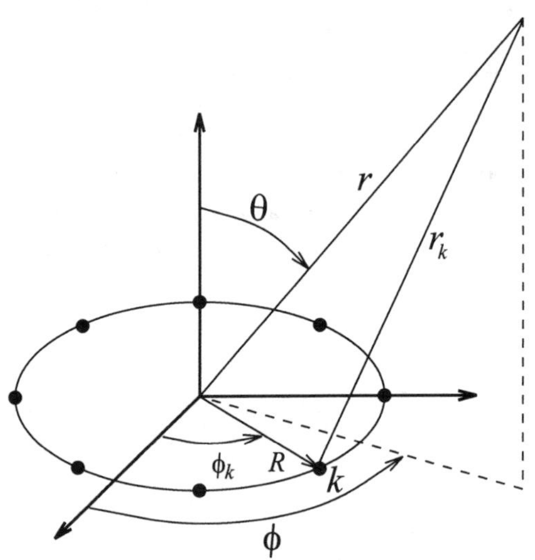

Figure 2.3 A circular array with equally spaced K elements.

It follows that the array factor for a circular array with K equally spaced elements is given as

$$F(\phi, \theta) = \sum_{k=0}^{K-1} A_k e^{j[\alpha_k - \kappa R \cos(\phi - \phi_k) \sin\theta]} \tag{2.13}$$

where $A_k e^{j\alpha_k}$ denotes the complex weight for the kth element. In order to have the main beam directed at the angle (ϕ_0, θ_0) in space, the phase of the weight for the kth element can be chosen to be

$$\alpha_k = \kappa R \cos(\phi_0 - \phi_k) \sin\theta_0 \tag{2.14}$$

The three-dimensional pattern of the array factor for an eight-element circular array with $R = 0.8710\lambda$ is shown in Figure 2.4.

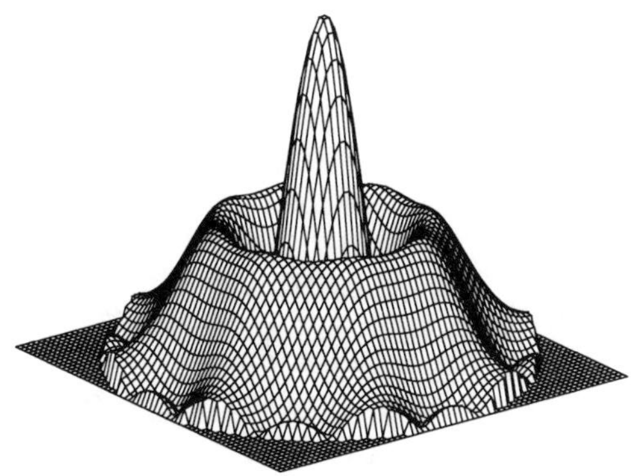

Figure 2.4 Three-dimensional beam pattern of an eight-element circular array with $R = 0.8710\lambda$.

In many applications, such as basestation antennas, the pattern in the $\theta = \pi/2$ plane is of interest. In this case, the array factor is given as

$$F(\phi) = \sum_{k=0}^{K-1} A_k e^{j[\alpha_k - \kappa R \cos(\phi - \phi_k)]} \tag{2.15}$$

An example of $F(\phi)$ for an eight-element circular array with $R = 0.6533\lambda$ is given in Figure 2.5, where the antenna beam is steered towards $\phi_0 = 0$.

One of the inherent characteristics of a circular array is the presence of high sidelobe levels in its beam pattern. For a circular array with equally spaced elements and uniform weighting, the lowest achievable peak sidelobe level is about 8 dB with respect to the main lobe. The sidelobe level is a function of θ_0 and ϕ_0 in addition to the physical parameters of the array. An excellent analysis on circular arrays was written by Ma [4].

2.1.4 Pattern Multiplication

So far, we have only considered arrays of isotropic antenna elements. An isotropic element can radiate or receive energy uniformly in all directions. The isotropic antenna is a mathematical fiction — it does not exist in practice. All practical antenna elements have nonuniform radiation patterns. Even a short dipole, which among realizable antennas, perhaps best approximates an isotropic antenna, has nulls located along its end points. Let us consider an array consisting of identical antenna elements that have radiation patterns decided by $f(\theta, \phi)$. The principle of pattern multiplication states that *the beam pattern of an array is the product of the*

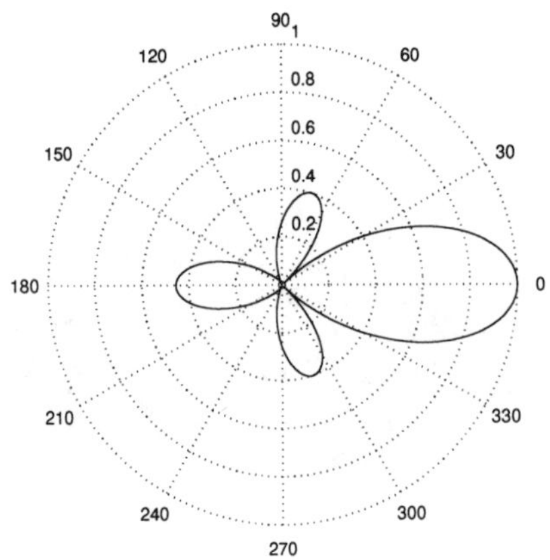

Figure 2.5 Two-dimensional beam pattern of an eight-element circular array with $R = 0.6533\lambda$, where the antenna beam is steered towards $\phi = 0$.

element pattern and the array factor. That is, the array beam pattern $G(\theta, \phi)$ is given by

$$G(\theta, \phi) = f(\theta, \phi)F(\theta, \phi) \tag{2.16}$$

where $F(\theta, \phi)$ is the array factor. The principle of pattern multiplication (2.16) is a very useful result. It shows how theorems relating to array design are independent of the particular antenna element used to form the array. It can also be used to determine the array factor of a complicated array that is composed of simple subarrays. For example, in the next section we will use the principle to determine the array factor of a planar array.

2.1.5 Planar Array

In addition to placing elements along a line to form a linear array, one can position them on a plane to form a planar array. In fact, the circular array is a special form of planar array, where the elements are placed along a circle that is usually located on a horizontal plane. Planar arrays provide additional variables which can be used to control and shape the array's beam pattern. The main beam of the array can be steered towards any point in its half space.

One of the common configurations of planar arrays is the rectangular array where the elements are placed along a rectangular grid, as shown in Figure 2.6. A rectangular array can be viewed as a linear array consisting of L identical elements

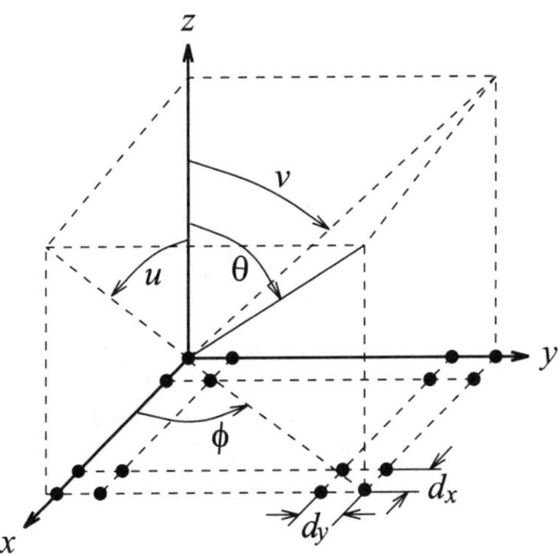

Figure 2.6 A rectangular planar array geometry.

each of which is a linear array with the array factor given by

$$F_1(u) = \sum_{k=0}^{K-1} A_k e^{j(\kappa k d_x \sin u + k\alpha)} \tag{2.17}$$

where $\sin u = \sin\theta\cos\phi$ and $\{A_k e^{jk\alpha}\}_{k=0}^{K-1}$ are the complex weights. The array factor for the L-element linear array is given as

$$F_2(v) = \sum_{l=0}^{L-1} B_l e^{j(\kappa l d_y \sin v + l\beta)} \tag{2.18}$$

where $\sin v = \sin\theta\sin\phi$ and $\{B_l e^{jl\beta}\}_{l=0}^{L-1}$ are the complex weights. According to the principle of pattern multiplication, the overall array factor for the rectangular array is then given by

$$F = F_1(u)F_2(v) \tag{2.19}$$

An example of $F(\theta,\phi)$ for an 8×8 rectangular array with $d_x = d_y = \lambda/2$ is given in Figure 2.7, where the antenna beam is steered towards $\theta_0 = 0$ and $\phi_0 = 0$. That is, $\alpha = -\kappa d \sin u_0$ and $\beta = \kappa d \sin v_0$

Another common configuration of planar arrays is the hexagon array, where the elements are placed along a triangular grid with an equal interelemental spacing d, as shown in Figure 2.8. Although the evaluation of the array factor for a hexagon planar array is not as simple as for a rectangular array, there are a number of ways to evaluate the array factor. A relatively simple way is to treat a hexagon array as

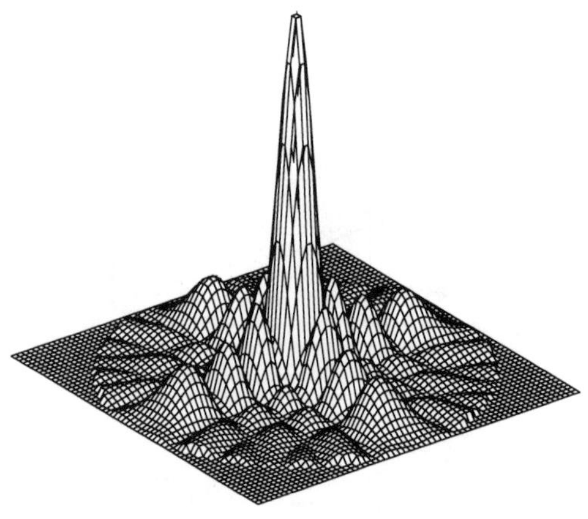

Figure 2.7 Three-dimensional beam pattern of an 8×8 rectangular array.

consisting of a single element at the center and a number of concentric six-element circular arrays of different radii, as shown in Figure 2.9. Thus, the overall array factor will be the sum of those for the circular arrays and the center element, which is given as

$$F(\theta,\phi) = A_0 + \sum_{k=1}^{K_h}\sum_{l=1}^{k}\sum_{m=0}^{5} A_{k,l,m} e^{j[\alpha_{k,l,m} - \kappa R_{k,l}\cos(\phi-\phi_{k,l,m})\sin\theta]} \tag{2.20}$$

where

$$R_{k,l} = d\sqrt{k^2 + (l-1)^2 - 2k(l-1)} \tag{2.21}$$

$$\phi_{k,l,m} = \arccos\left[\frac{R_{k,l}^2 + dk^2 - d(l-1)^2}{2R_{k,l}dk}\right] + m\frac{\pi}{3} \tag{2.22}$$

and K_h is the number of hexagons. For example, in the case of the array shown in Figure 2.8, the value of K_h is equal to 3. An example of $F(\theta,\phi)$ for this particular array is shown in Figure 2.10, where the antenna beam is steered towards $\theta_0 = 0$ and $\phi_0 = 0$.

2.2 ANALOG BEAMFORMING

The term beamforming relates to the function performed by a device or apparatus in which energy radiated by an aperture antenna is focused along a specific direction in space. The objective is either to preferentially receive a signal from that direction or to preferentially transmit a signal in that direction. For example, in a

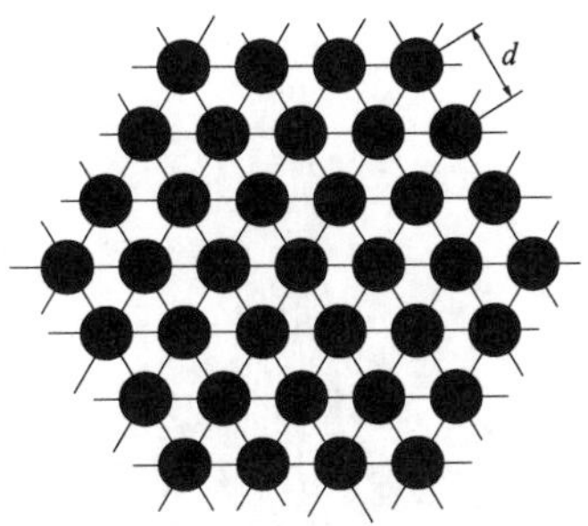

Figure 2.8 A hexagon planar array geometry.

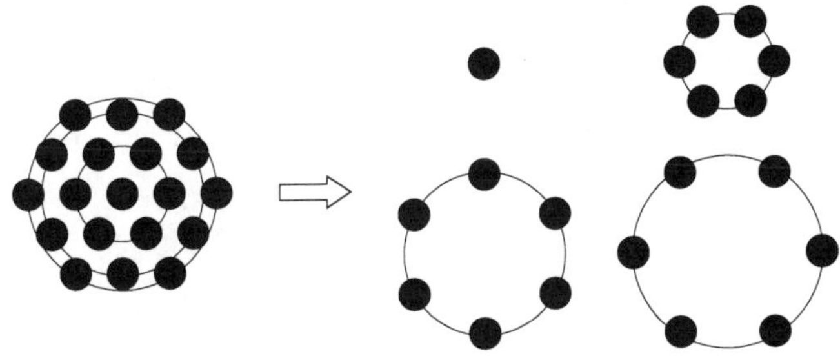

Figure 2.9 A hexagon planar array can be considered as consisting of a number of concentric six-element circular arrays of different radius.

parabolic antenna system, the dish is the beamforming network in that it takes the energy that lies within the aperture formed by the perimeter of the dish and focuses it onto the antenna feed. The dish and feed operate as a spatial integrator. Energy from a far-field source, which is assumed to be aligned with the antenna's preferred direction, arrives at the feed temporarily aligned and is thereby summed coherently. In general, sources in other directions arrive at the feed unaligned and add incoherently. For this reason, beamforming is often referred to as spatial filtering.

Spatial filtering may also be carried out using antenna arrays. In fact, an array can be considered as a sampled aperture. When an array is illuminated by a source, samples of the source's wavefront are recorded at the location of the antenna elements. The outputs from elements can be subjected to various forms of signal processing, wherein phase or amplitude adjustments are made to produce outputs that can provide concurrent angular information for signals arriving in several different directions in space. When the outputs of the elements of an array are

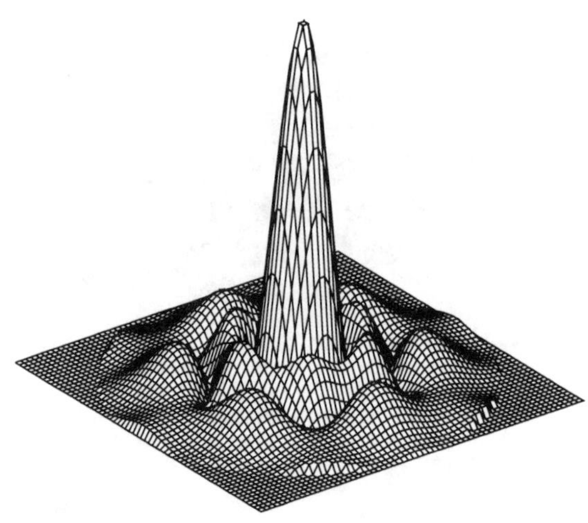

Figure 2.10 Three-dimensional beam pattern of a 37-element hexagon array.

combined via some passive phasing network, the phasing will usually arrange for the output of all the elements to add coherently for a given direction. If information is desired regarding signals arriving from a different region in space, another phasing network would have to be implemented. The network that controls the phases and amplitudes of the excitation current is usually called the beamforming network. If beamforming is carried out at RF, the analog beamforming network usually consists of devices that change the phase and power of the signals. Figure 2.11 gives an example of an RF beamformer that is designed to form only one beam. The beamforming network can be implemented using microwave lenses, waveguides, transmission lines, printed microwave circuits, and hybrids. Figure 2.12 shows a four-element microstrip antenna array with a beamforming network. This simple topology is capable of forming only one beam. Furthermore, it only provides for uniform weighting, wherein the array pattern is given by the sine function.

It is sometimes desirable to form multiple beams that are offset by finite angles from each other. The design of a multiple-beam beamforming network is much more complicated than that of a single-beam beamforming network. A multiple-beam beamforming network is known as a beamforming matrix. The best known example is given by the Butler matrix [5]. In a beamforming matrix, an array of hybrid junctions and fixed-phase shifters are used to achieve the desired results. For example, a Butler beamforming matrix for a four-element array is shown in Figure 2.13(a). This matrix uses four 90° phase-lag hybrid junctions with the transmission properties shown in Figure 2.13(b) and two 45° fixed-phase shifters.

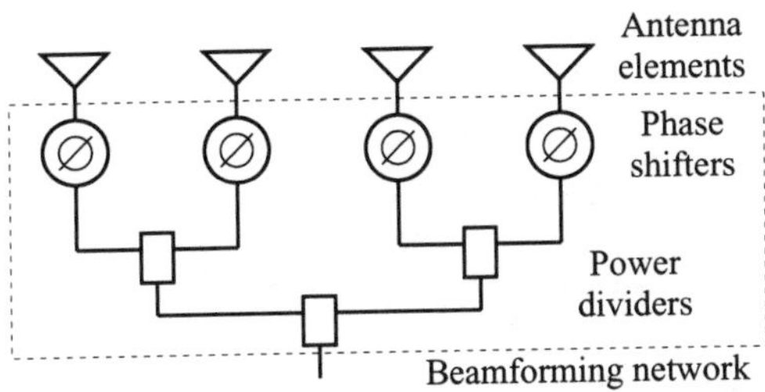

Figure 2.11 An analog beamforming network consists of devices such as phase shifters and power dividers which are used to adjust the amplitudes and phases of the elemental signals in such a way as to form a desired beam.

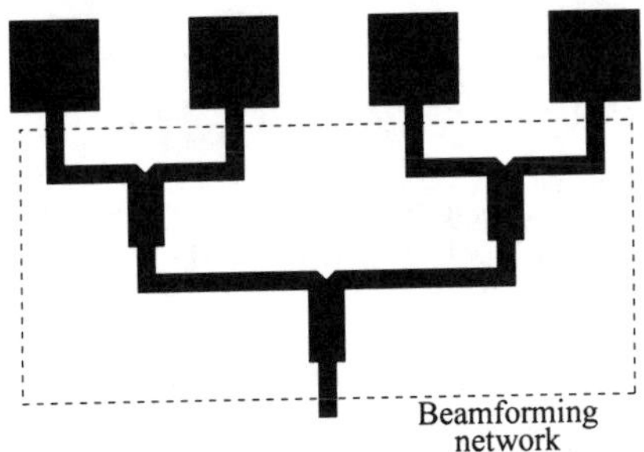

Figure 2.12 A four-element microstrip antenna array with a beamforming network of uniform phase and amplitude weighting.

By tracing the signal from the four ports to the array elements, one should be able to verify the aperture relative phase distribution corresponding to the individual ports of a four-port Butler matrix, shown in Table 2.1. If the elemental spacing is $\lambda/2$, the system produces four beams as shown in Figure 2.14. Although, these four beams overlap, they are mutually orthogonal.

Butler matrix beamforming is similar to the *fast Fourier transform* (FFT) process. In fact, they have a 1 : 1 equivalence. Surprisingly, the Butler matrix was developed before the FFT. However, there is an important difference between them: a Butler matrix processes signals in the analog domain, whereas the FFT processes signals in the digital domain.

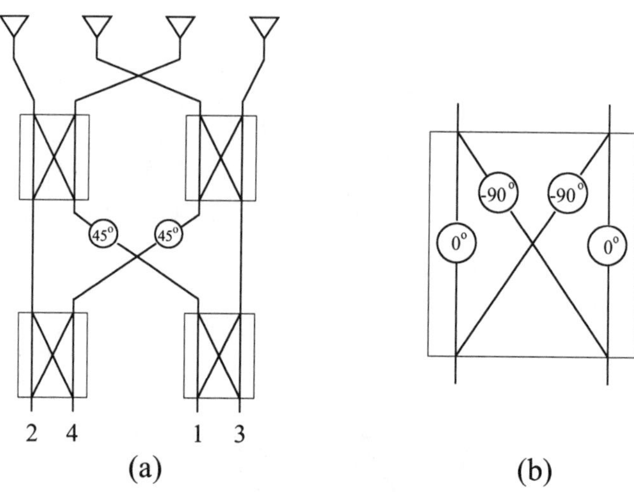

Figure 2.13 A Butler beamforming matrix for a four-element array: (a) 4×4 Butler matrix; (b) a hybrid used in the matrix.

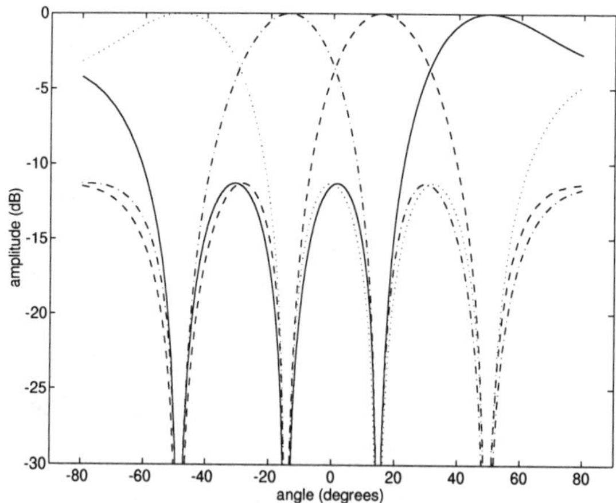

Figure 2.14 Four mutually orthogonal overlapped beams produced by the Butler beamforming matrix.

Table 2.1
Array Aperture Phase Distribution

	Phase Distribution			
Port 1	0°	-135°	-270°	-405°
Port 2	0°	-45°	-90°	-135°
Port 3	0°	45°	90°	135°
Port 4	0°	135°	270°	405°

2.3 PHASED ARRAYS

It is often desirable to electronically scan the beam of an antenna. This can be accomplished by changing the phases of the signals at the antenna elements. If only the phases are changed, with the amplitude weights remaining fixed as the beam is steered, the array is commonly known as a phased array. As shown in Figure 2.15, a phased array consists simply of antenna elements, each of which is connected to a phase shifter, and a power combiner for adding together the signals from the antenna elements. The phase shifters control either the phase of the excitation current or the phase of the received signals. When the all signals are combined, a beam is formed in the desired direction. That is, on transmit a beam is formed in space, and on receive the signals from the antenna elements add coherently if the signals are received from the correct region of space. A beamforming network is used to either distribute the signal from the transmitter to the elements or combine the signals from the elements to form a single signal path to the receiver. The network may also be used to provide the required aperture distribution for beam shaping and side lobe control. The phase shifters can be classified into two types: continuously variable phase shifters and digitally controlled phase shifters. Since these components operate at RF, they have a small error tolerance. Thus, the design and manufacturing of these devices can be costly.

Alternatively, beamforming can be carried out at intermediate frequencies (IF). The beamforming network can be implemented using resistors, hybrid circuits, and tapped delay lines, which are constructed using lumped circuits. IF beamforming may be more convenient in many ways, since it may be performed after amplification has taken place so that the losses in the beamforming network are less important. However, it requires that each element must have its own RF-to-IF receiver.

If multiple beams are required, multiple-beam beamformers must be used, which distribute the signal energy to all the formed beams. However, there are a number

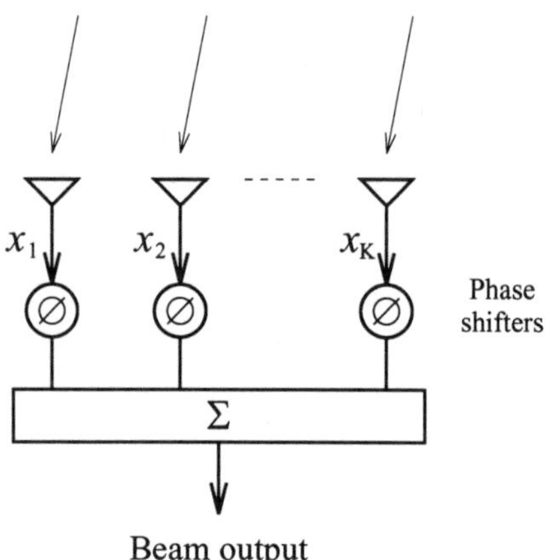

Figure 2.15 A linear phased array.

of drawbacks with multiple-beam beamformers. First, it is difficult to reconfigure the beamformer (i.e., to expand or modify its characteristics). Most multiple-beam beamformers can only produce fixed beams. Second, as the number of beams is increased, the SNR of channels being carried by the individual beams decreases. This comes about because additional noise is introduced due to the increased number of RF and IF components that must be used to increase the beamformer's capacity. Third, the separation between the multiple beams cannot be any less than that for orthogonal beams without a considerable decrease in system SNR.

2.4 DIGITAL BEAMFORMING

The early ideas that form the foundations of digital beamforming were first developed by workers in sonar [6] and radar systems [7]. DBF is a marriage between antenna technology and digital technology. An antenna can be considered to be a device that converts spatiotemporal signals into strictly temporal signals, thereby making them available to a wide variety of signal processing techniques. In this way all of the desired information that is being carried by these signals can be extracted. An optimum antenna is one that carries out the conversion of the signals that arrive at its face without introducing any distortions to the signals. It is for this reason that a digital beamforming antenna might be considered to be an optimum antenna. From a conceptual point of view, its sampled outputs represent all of the data arriving at the antenna aperture. No information is destroyed, at least, not until the processing begins, and any compromises that are made in the processing stages can be noted and estimates made of the divergence of the actual system from the ideal.

A major advantage of digital beamforming lies in the fact that once the RF information is captured in the form of a digital stream, the way is clear for the application of a multitude of digital signal processing techniques and algorithms to the spatial domain data. Digital beamforming is based on the conversion of the RF signal at each antenna elements into two streams of binary baseband signals representing I and Q channels. The digital baseband signals then represent the amplitudes and phases of signals received at each element of the array. The process of beamforming implies weighting these digital signals, thereby adjusting their amplitudes and phases such that when added together they form the desired beam. The beam decided by the array factor (2.11) is actually formed in the digital domain (i.e., within the computer). It does not have the same direct physical meaning as it does in the case of analog beamforming. Nevertheless, its effect on the performance of a communications link is every bit as real as in the case of analog beamforming. The key to this technology is the accurate translation of the analog signal into the digital regime. This is accomplished using complete heterodyne receivers, which must be closely matched in amplitude and phase. The matching need not be carried out by making adjustments to the hardware, but rather by applying a calibration process that adjusts the values of the data stream prior to beamforming. The calibration is carried out in software. The receivers are required to perform frequency down-conversion, filtering, and amplification to a power level commensurate with the input requirements for the ADCs. In addition to the calibration being carried out using software, it is expected increasingly that more and more of the receiver functions will be implemented using software. Eventually, one would expect that the receiver would be built using software rather than hardware. The main advantage to be gained from digital beamforming is the greatly added flexibility without any attendant degradation in SNR. In many ways it can be considered to be the ultimate antenna, in that all of the information arriving at the antenna face is captured by digital streams. All of this information is available for processing in the beamformer. There are many possible configurations for performing the digital processing. Beamforming instructions can be driven by software to produce different types of beams, such as scanned beams, multiple beams, shaped beams, or beams with steered nulls.

Figure 2.16 depicts a simple structure that can be used for beamforming. The beamformer, which carries out sampling of the propagated wave field, is typically used for processing narrowband signals. The output at time n, $y(n)$, is given by a linear combination of the data at the K sensors at time n:

$$y_n(\theta) = \sum_{k=0}^{K-1} w_k^* x_k(n) \tag{2.23}$$

where * represents a complex conjugate, x_k is the signal from the kth element of the array, and w_k is the weight applied to x_k. Following the established convention, we multiply the data by conjugates of the weights. This convention helps to simplify the notation. In vector form, (2.23) can be written as

$$y_n(\theta) = \boldsymbol{w}^H \boldsymbol{x}(n) \tag{2.24}$$

where the superscript H represents Hermitian transpose. If $x_k(n) = V_k$ and $w_k = e^{-j\kappa kd \sin\theta}$, the output of y is equal to F in (2.11); that is, $y_n(\theta) = F(\theta)$. With the digital beamformer, however, one is able to do a lot more than just the evaluation of the array factor. In fact, one is able to change the value of $\boldsymbol{w}$ to point the beam in any wanted direction and to manipulate its shape to optimize system performance. Therefore, the flexibility of DBF allows the full implementation of adaptive beamforming, which will be discussed in the next chapter.

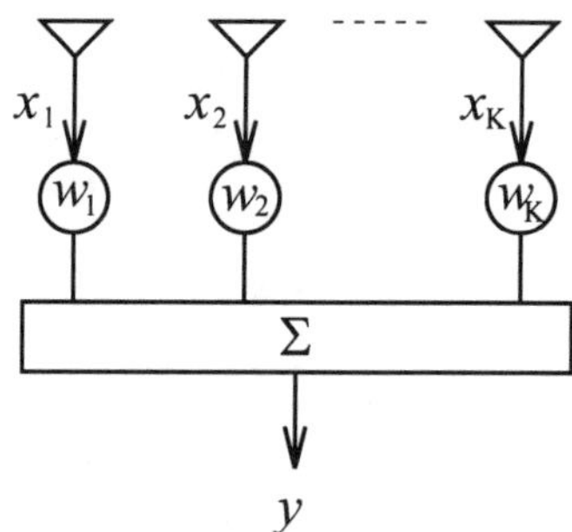

Figure 2.16 A beamformer forms a linear combination of the sensor outputs, which are first multiplied by complex weights and then summed together.

2.4.1 Element-Space Beamforming

The process represented by (2.23) is referred to as element-space beamforming, where the data signals x_k, $k = 0, \cdots, K-1$, from the array elements are directly multiplied by a set of weights to form a beam at a desired angle. By multiplying the data signals by different sets of weights, it is possible to form a set of beams with pointing angles directed anywhere in the field defined by the elements used in the array. Each beamformer creates an independent beam by applying independent weight to the array signals:

$$y(\theta_i) = \sum_{k=0}^{K-1} w_k^{i*} x_k \tag{2.25}$$

where

$y(\theta_i)$ = output of the beamformer;
x_k = sample from the kth array element;
w_k^i = weights for forming beam at angle θ_i.

The setup for generating an arbitrary number of simultaneous beams from K antenna elements is shown in Figure 2.17. By selecting appropriate values for the weighting vectors, one can implement beam steering, adaptive nulling, and beamshaping.

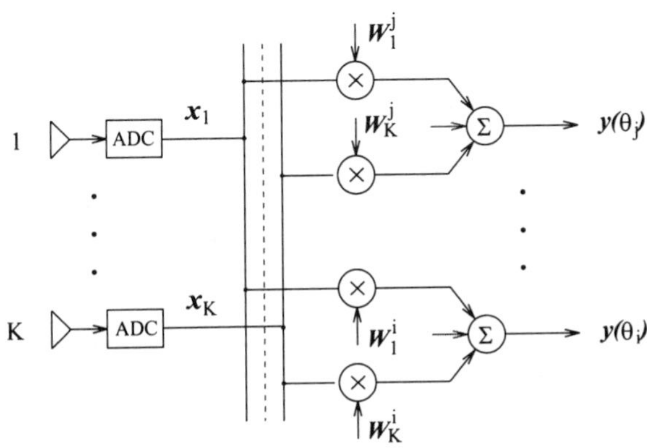

Figure 2.17 Element-space digital beamformer for simultaneously generating L beams.

2.4.2 Beam-Space Beamforming

Rather than directly weighting the outputs from the array elements, they can be first processed by a multiple-beam beamformer to form a suite of orthogonal beams. The output of each beam can then be weighted and the result combined to produce a desired output. This process is often referred to as beam-space beamforming. The required multiple beamformer usually produces orthogonal beams. For example, the beamformer can be implemented by using the FFT. For a K-element linear array,

K overlapped, orthogonal beams can be formed; that is,

$$v(\xi_l) = \sum_{k=1}^{K} x_k e^{-j2\pi k\xi_l/K}, \qquad for \ \ l = 1, \ \cdots \ K \tag{2.26}$$

where $\theta_l = \sin^{-1}(l\lambda/Kd)$. Because of the fixed nature of these unweighted output beams (i.e., discrete ξ_l), $v(\xi_l)$, individual beam control requires the following:

1. Interpolation between beams in order to fine-steer the resultant beam;
2. Linear combination of the output beams to synthesize a shaped beam or a low sidelobe pattern;
3. Linear combination of a selected set of beams to create nulls in the direction of the interfering sources.

In other words, beam-space beamforming requires a set of beam-space combiners to generate weighted outputs. Figure 2.18 shows the implementation of a weighted FFT-based beamformer. The digital signal streams from the antenna elements are fed to the FFT processor, which generates K simultaneous orthogonal beams. The role of the beam select function is to choose a subset of these orthogonal beams that are to be weighted to form a desired output. For instance, the ith desired output may happen to be the combination of the weighted k^{th} and $(k+2)^{\text{th}}$ beams. That is,

$$y_i = w_1^i v(\xi_k) + w_2^i v(\xi_{k+2}) = \sum_{m=1}^{M_i} w_m^i v(\xi_{i(m)}) \tag{2.27}$$

where $i(m)$ is the selected beam index (e.g., $i(1) = k$ and $i(2) = k+2$) and M_i is the number of orthogonal beams that are required to form the ith desired beam.

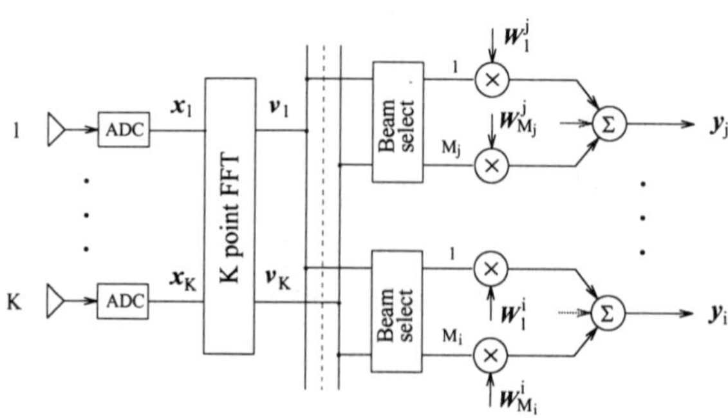

Figure 2.18 Beam-space digital beamformer for simultaneously generating multiple beams.

2.4.3 Two-Dimensional Beamforming

In mobile satellite communications applications, the antenna arrays are usually two-dimensional. Tthe digital beamforming concepts, techniques, and algorithms for a linear antenna array can be easily and naturally extended to a two-dimensional planar array.

For a K-by-L rectangular planar array, the output of the beamformer at time n, $y_n(\theta, \phi)$, is given by

$$y_n(\theta, \phi) = \sum_{k=0}^{K-1} \sum_{l=0}^{L-1} w_{k,l}^* x_{k,l}(n) \tag{2.28}$$

which can be written as

$$y_n(\theta, \phi) = \boldsymbol{w}^H \boldsymbol{x}(n) \tag{2.29}$$

where

$$\boldsymbol{w} = [w_{0,0},\ w_{1,0},\ \cdots,\ w_{K-1,0},\ w_{0,1},\ \cdots,\ w_{K-1,L-1}]^T \tag{2.30}$$

and

$$\boldsymbol{x}(n) = [x_{0,0}(n),\ x_{1,0}(n),\ \cdots, x_{K-1,0}(n),\ x_{0,1}(n),\ \cdots,\ x_{K-1,L-1}(n)]^T \tag{2.31}$$

In a similar way, for a hexagon planar array, the output of the beamformer at time $n, y_n(\theta, \phi)$, which is given by

$$y_n(\theta, \phi) = w_0^* x_0 + \sum_{k=1}^{K_h} \sum_{l=1}^{k} \sum_{m=0}^{5} w_{k,l,m}^* x_{k,l,m}(n) \tag{2.32}$$

can be expressed as

$$y_n(\theta, \phi) = \boldsymbol{w}^H \boldsymbol{x}(n) \tag{2.33}$$

where

$$\boldsymbol{w} = [w_0,\ w_{1,1,1},\ w_{2,1,1},\ \cdots,\ w_{K_h,1,1},\ w_{2,2,1},\ \cdots,\ w_{K_h,K_h-1,5}]^T \tag{2.34}$$

and

$$\boldsymbol{x}(n) = [x,0(n),\ x_{1,1,1}(n),\ x2,1,1,\ \cdots, x_{K_h,1,1}(n),\ x_{2,2,1}(n),\ \cdots,\ x_{K_h,Kh-1,5}(n)]^T \tag{2.35}$$

The reason why we express the beamformer output $y_n(\theta, \phi)$ for a planar array as an inner product of two vectors is to make it mathematically equivalent to (2.23). This allows the beamforming techniques and algorithms that apply to one-dimensional array to be applicable to planar arrays, as well.

References

[1] Y. T. Lo and S. W. Lee, eds., *Antenna Handbook, Theory, Applications and Design*, Van Nostrand Reinhold Company, New York, 1988.

[2] A. W. Rudge, K. Milne, A. D. Olver, and P. Knight, eds., *The Handbook of Antenna Design*, Peter Peregrinus, London, 1986.

[3] J. D. Kraus, *Antennas*, McGraw-Hill, New York, 1988.

[4] M. T. Ma, *Theory and Application of Antenna Arrays*, John Wiley & Sons, New York, 1974.

[5] J. L. Butler, "Digital, matrix, and intermediate frequency scanning," in R. C. Hansen, ed., *Microwave Scanning Arrays*, Academic Press, New York, 1966.

[6] T. E. Curtis, "Digital beam forming for sonar system," *IEE Proc. Pt. F*, vol. 127, pp. 257–265, Aug. 1980.

[7] P. Barton, "Digital beamforming for radar," *IEE Proc. Pt. F*, vol. 127, pp. 266–277, Aug. 1980.

Chapter 3

Adaptive Beamforming

An adaptive beamformer is a device that is able to separate signals collocated in the frequency band but separated in the spatial domain. This provides a means for separating a desired signal from interfering signals. An adaptive beamformer is able to automatically optimize the array pattern by adjusting the elemental control weights until a prescribed objective function is satisfied. The means by which the optimization is achieved is specified by an algorithm designed for that purpose. These devices use far more of the information available at the antenna aperture than does a conventional beamformer.

Traditionally, adaptive beamforming has been employed primarily in sonar and radar systems. It started with the invention of the intermediate frequency sidelobe conceler (SLC) in 1959 by Howells [1]. The concept of a fully adaptive array was developed in 1965 by Applebaum [2]. He derived the control law governing its operation. This algorithm is based on the general problem of maximization of SNR at the array output. The SLC was included as a special case in Applebaum's work. Another independent approach to adaptivity uses the least mean squares (LMS) error algorithm, which was invented by Widrow and Hoff [3]. Despite its simplicity, the LMS algorithm is capable of achieving satisfactory performance under the right set of conditions. The LMS algorithm was further developed with the introduction of constraints [4,5]. Constraints are used to ensure that the desired signals are not filtered out along with the unwanted signals. With a constraint in place, the optimization process proceeds as before, but now, for example, with the antenna gain maintained constant in the desired look direction. Although Applebaum's maximum SNR algorithm and Widrow's LMS algorithm were discovered independently and were developed using two different approaches, they are basically similar. For stationary signals, both algorithms converge to the optimum Wiener solution [6].

A different technique for solving the adaptive beamforming problem was proposed in 1969 by Capon [7]. His algorithm leads to an adaptive beamformer with a minimum-variance distortionless response (MVDR). This has also been referred to by some writers as the maximum likelihood method (MLM), because the algorithm maximizes the likelihood function of the input signal vector. It is also one of the earliest adaptive beamforming techniques that offers the ability to resolve signals that are separated by a fraction of an antenna beamwidth. In 1974, Reed and his coworkers showed that fast adaptivity is achieved by using the sample-matrix inversion (SMI) technique [8]. Using this technique, the adaptive weights can be computed directly. Unlike the maximum SNR algorithm and the LMS algorithm, which may suffer from slow convergence, if the value of the eigenvalue spread of the sample covariance matrix (SCM) is relatively large, the performance of the SMI scheme is almost independent of the value of the eigenvalue spread.

Adaptive beamforming has been a subject of considerable interest for more than three decades. It has evolved into a mature technology. There are a number of textbooks and tutorial papers devoted to general concepts and applications of adaptive beamforming [9–13]. In addition, numerous papers and articles on specific techniques and applications can be found in the open literature. Thus, here we will only present some basic adaptive beamforming concepts and a number of representative adaptive techniques that can be used in wireless communications.

3.1 BASIC CONCEPTS

The procedure used for steering and modifying an array's beam pattern in order to enhance the reception of a desired signal, while simultaneously suppressing interfering signals through complex weight selection, is illustrated by the following example. Consider the array shown in Figure 3.1, which consists of two omnidirectional antennas with $\frac{\lambda_0}{2}$ spacing. The desired signal, $S(t)$, arrives from the boresight direction ($\theta_S = 0$), and the interference signal, $I(t)$, arrives from the angle ($\theta_I = \pi/6$) radians. Both signals have the same frequency f_0. The signal from each element is multiplied by a variable complex weight, and the weighted signals are then summed to form the array output. The array output due to the desired signal is

$$y_d(t) = Ae^{j2\pi f_0 t}(w_1 + w_2) \tag{3.1}$$

For $y_d(t)$ to be equal to $S(t)$, it is necessary that

$$\left.\begin{aligned} \Re[w_1] + \Re[w_2] = 1 \\ \Im[w_1] + \Im[w_2] = 0 \end{aligned}\right\} \tag{3.2}$$

where $\Re[]$ and $\Im[]$ operate on the real and imaginary values, respectively. The incident interference signal arrives at element 2 with a phase lead with respect to element 1 of value $2\pi\frac{1}{\lambda_0}d\sin(\pi/6) = \pi/2$. Consequently, the array output due to the interference is

$$y_I(t) = Ne^{j2\pi f_0 t}w_1 + Ne^{j(2\pi f_0 t+\pi/2)}w_2 \tag{3.3}$$

For the array interference response to be zero, it is necessary that

$$\left.\begin{aligned} \Re[w_1] + \Re[jw_2] = 0 \\ \Im[w_1] + \Im[jw_2] = 0 \end{aligned}\right\} \tag{3.4}$$

Simultaneous solution of (3.2) and (3.4) yields

$$w_1 = \frac{1}{2} - j\frac{1}{2}, w_2 = \frac{1}{2} + j\frac{1}{2}$$

With these weights, the array will accept the desired signal while simultaneously rejecting the interference.

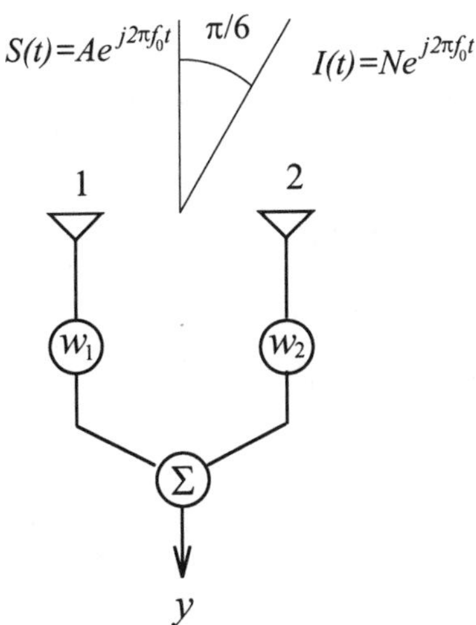

Figure 3.1 Two-element array for interference suppression.

The method used in the above example exploits the fact that there is only one directional interference source and uses the *a priori* information concerning the frequency and the directions of both of the signals. A more practical processor should not require such detailed *a priori* information about the location, number, and nature of the signal sources. Nevertheless, this example has demonstrated that a system consisting of an array, which is configured with complex weights, provides countless possibilities for realizing array system objectives. We need only develop a practical adaptive processor for carrying out the complex weight adjustment.

A generic adaptive beamforming system is shown in Figure 3.2. The choice of the weight vector $\boldsymbol{w}$ is based on the statistics of the signal vector $\boldsymbol{x}(t)$ received at the array. Basically, the objective is to optimize the beamformer response with respect to a prescribed criterion, so that the output $y(t)$ contains minimal contribution from noise and interference. There are a number of criteria for choosing the optimum weights, which will be discussed in the next section.

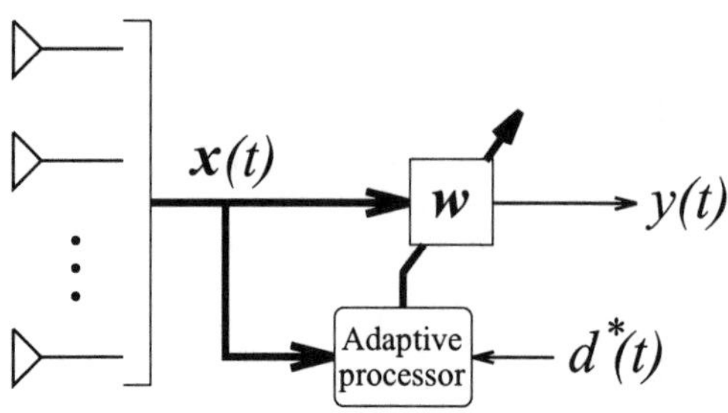

Figure 3.2 A generic adaptive beamforming system.

3.2 CRITERIA FOR OPTIMAL WEIGHTS

3.2.1 Minimum Mean-Square Error

Let us consider a uniformly spaced linear array, as shown in Figure 2.1, which operates in a signal environment where there is a desired communication signal $s(t)$, as well as N_u interfering signals $\{u_i(t)\}_{i=1}^{N_u}$. Let us further assume that the desired signal arrives at the array with a spatial angle θ_0 and the ith interfering signal arrives with an angle θ_i. The array output is represented by

$$\boldsymbol{x}(t) = s(t)\boldsymbol{v} + \boldsymbol{u} = \boldsymbol{s} + \boldsymbol{u} \tag{3.5}$$

where $\boldsymbol{v}$ is the array propagation vector (see Section 2.1.2) for the desired signal,

$$\boldsymbol{v}^T = [1, \ e^{j\kappa d \sin\theta_0} \ \cdots \ e^{j\kappa(K-1)d\sin\theta_0}] \tag{3.6}$$

$\boldsymbol{u}$ represents the sum of all the interfering signal vectors,

$$\boldsymbol{u} = \sum_{i=1}^{N_u} u_i(t)\boldsymbol{\eta}_i \tag{3.7}$$

and $\boldsymbol{\eta}_i$ is the array propagation vector for the ith interfering signal,

$$\boldsymbol{\eta}_i^T = [1, \ e^{j\kappa d \sin\theta_i} \ \cdots \ e^{j\kappa(K-1)d\sin\theta_i}] \tag{3.8}$$

If the desired signal $s(t)$ is known, one may choose to minimize the error between the beamformer output $\boldsymbol{w}^H\boldsymbol{x}(t)$ and the desired signal. Of course, knowledge of the desired signal eliminates the need for beamforming. However, for many applications, characteristics of the desired signal may be known with sufficient detail to generate a signal $d^*(t)$ that closely represents it, or at least correlates with the desired signal to a certain extent. This signal is called a reference signal [14]. It should be noted that we express the reference signal as a complex conjugate only for mathematical convenience. It would not make any difference in the final result. The weights are chosen to minimize the mean-square error (MSE) between the beamformer output and the reference signal:

$$\epsilon^2(t) = [d^*(t) - \boldsymbol{w}^H\boldsymbol{x}(t)]^2 \tag{3.9}$$

Taking the expected values of both sides of (3.9) and carrying out some basic algebraic manipulation, we have

$$E\{\epsilon^2(t)\} = E\{d^2(t)\} - 2\boldsymbol{w}^H\boldsymbol{r} + \boldsymbol{w}^H\boldsymbol{R}\boldsymbol{w} \tag{3.10}$$

where $\boldsymbol{r} = E\{d^*(t)\boldsymbol{x}(t)\}$ and $\boldsymbol{R} = E\{\boldsymbol{x}(t)\boldsymbol{x}^H(t)\}$. $\boldsymbol{R}$ is usually referred to as the covariance matrix. The minimum (MSE) is given by setting the gradient vector of (3.10) with respect to $\boldsymbol{w}$ equal to zero:

$$\begin{aligned}\nabla_{\boldsymbol{w}}(E\{\epsilon^2(t)\}) &= -2\boldsymbol{r} + 2\boldsymbol{R}\boldsymbol{w} \\ &= 0\end{aligned} \tag{3.11}$$

It follows that the solution is

$$\boldsymbol{w}_{\text{opt}} = \boldsymbol{R}^{-1}\boldsymbol{r} \tag{3.12}$$

which is referred to as the Wiener-Hopf equation or the optimum Wiener solution. If $s(t) = d^*(t), \boldsymbol{r} = E\{d^2(t)\}\boldsymbol{v}$. Let us further express $\boldsymbol{R} = E\{d^2(t)\}\boldsymbol{v}\boldsymbol{v}^H + \boldsymbol{R}_u$, where $\boldsymbol{R}_u = E\{\boldsymbol{u}\boldsymbol{u}^H\}$ and apply *Woodbury's Identity* to $\boldsymbol{R}^{-1}$ and we have

$$\boldsymbol{R}^{-1} = \left[\frac{1}{1 + E\{d^2(t)\}\boldsymbol{v}^H\boldsymbol{R}_u^{-1}\boldsymbol{v}}\right]\boldsymbol{R}_u^{-1} \tag{3.13}$$

Therefore, the Wiener solution can be generalized as

$$\boldsymbol{w}_{\text{opt}} = \beta\boldsymbol{R}_u^{-1}\boldsymbol{v} \tag{3.14}$$

where β is a scalar coefficient. In the case of minimum MSE,

$$\beta = \frac{E\{d^2(t)\}}{1 + E\{d^2(t)\}\boldsymbol{v}^H\boldsymbol{R}_u^{-1}\boldsymbol{v}} \tag{3.15}$$

3.2.2 Maximum Signal-to-Interference Ratio

The weights can be chosen to directly maximize the signal-to-interference ratio (SIR). Assuming that $\boldsymbol{R}_s = E\{\boldsymbol{s}\boldsymbol{s}^H\}$ and $\boldsymbol{R}_u = E\{\boldsymbol{u}\boldsymbol{u}^H\}$ are known, we may choose to maximize the ratio of the output signal power σ_s^2 and the total interfering signal power σ_u^2. The output signal power may be written as

$$\sigma_s^2 = E\{|\boldsymbol{w}^H \boldsymbol{s}|^2\} = \boldsymbol{w}^H \boldsymbol{R}_s \boldsymbol{w} \tag{3.16}$$

and the output noise power is

$$\sigma_u^2 = E\{|\boldsymbol{w}^H \boldsymbol{u}|^2\} = \boldsymbol{w}^H \boldsymbol{R}_u \boldsymbol{w} \tag{3.17}$$

Therefore, the (SIR) is given as

$$\text{SIR} = \frac{\sigma_s^2}{\sigma_u^2} = \frac{\boldsymbol{w}^H \boldsymbol{R}_s \boldsymbol{w}}{\boldsymbol{w}^H \boldsymbol{R}_u \boldsymbol{w}} \tag{3.18}$$

Taking the derivative of (3.18) with respect to $\boldsymbol{w}$ and setting it to zero, we obtain

$$\boldsymbol{R}_s \boldsymbol{w} = \frac{\boldsymbol{w}^H \boldsymbol{R}_s \boldsymbol{w}}{\boldsymbol{w}^H \boldsymbol{R}_u \boldsymbol{w}} \boldsymbol{R}_u \boldsymbol{w} \tag{3.19}$$

which appears to be a joint eigenproblem. The value of the $\frac{\boldsymbol{w}^H \boldsymbol{R}_s \boldsymbol{w}}{\boldsymbol{w}^H \boldsymbol{R}_u \boldsymbol{w}}$ is bounded by the minimum and maximum eigenvalues of the symmetric matrix $\boldsymbol{R}_u^{-1} \boldsymbol{R}_s$. The maximum eigenvalue λ_{max} satisfying

$$\boldsymbol{R}_u^{-1} \boldsymbol{R}_s \boldsymbol{w} = \lambda_{\text{max}} \boldsymbol{w} \tag{3.20}$$

is the optimum value of (SIR); (i.e., SIR $= \lambda_{\text{max}}$). Corresponding to this value, there is a unique eigenvector, $\boldsymbol{w}_{\text{opt}}$, which represents the optimum weights. Therefore,

$$\boldsymbol{R}_s \boldsymbol{w}_{\text{opt}} = \text{SIR} \boldsymbol{R}_u \boldsymbol{w}_{\text{opt}} \tag{3.21}$$

Noting that $\boldsymbol{R}_s = E\{d^2(t)\} \boldsymbol{v}\boldsymbol{v}^H$, we obtain

$$\boldsymbol{w}_{\text{opt}} = \beta \boldsymbol{R}_u^{-1} \boldsymbol{v} \tag{3.22}$$

where

$$\beta = \frac{E\{d^2(t)\}}{\text{SIR}} \boldsymbol{v}^H \boldsymbol{w}_{\text{opt}} \tag{3.23}$$

That is, the maximum (SIR) criterion can also be expressed in terms of the Wiener solution.

3.2.3 Minimum Variance

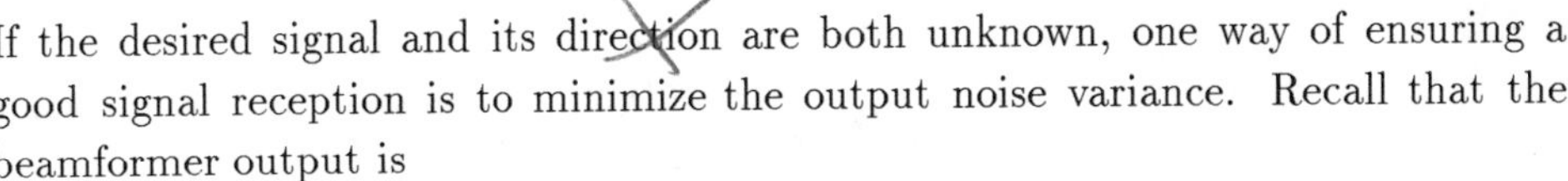

If the desired signal and its direction are both unknown, one way of ensuring a good signal reception is to minimize the output noise variance. Recall that the beamformer output is

$$\begin{aligned} y(t) &= \boldsymbol{w}^H \boldsymbol{x} \\ &= \boldsymbol{w}^H \boldsymbol{s} + \boldsymbol{w}^H \boldsymbol{u} \end{aligned} \tag{3.24}$$

To ensure that the desired signal is passed with a specific gain and phase, a constraint may be used so that the response of the beamformer to the desired signal is

$$\boldsymbol{w}^H \boldsymbol{v} = g \tag{3.25}$$

Minimization of contributions to the output due to interference is accomplished by choosing the weights to minimize the variance of the output power

$$\begin{aligned} \mathrm{Var}\{y\} &= \boldsymbol{w}^H \boldsymbol{R} \boldsymbol{w} \\ &= \boldsymbol{w}^H \boldsymbol{R}_s \boldsymbol{w} + \boldsymbol{w}^H \boldsymbol{R}_u \boldsymbol{w} \end{aligned} \tag{3.26}$$

subject to the constraint defined in (3.25). This is equivalent to minimizing the quantity $\boldsymbol{w}^H \boldsymbol{R}_u \boldsymbol{w}$. Using the method of Lagrange, we have

$$\nabla_{\boldsymbol{w}}\left(\frac{1}{2}\boldsymbol{w}^H \boldsymbol{R}_u \boldsymbol{w} + \beta[1 - \boldsymbol{w}^H \boldsymbol{v}]\right) = \boldsymbol{R}_u \boldsymbol{w} - \beta \boldsymbol{v} \tag{3.27}$$

so that

$$\boldsymbol{w}_{\text{opt}} = \beta \boldsymbol{R}_u^{-1} \boldsymbol{v} \tag{3.28}$$

where

$$\beta = \frac{g}{\boldsymbol{v}^H \boldsymbol{R}_u^{-1} \boldsymbol{v}} \tag{3.29}$$

That is, solution (3.28), which is derived from the minimum variance criterion, is also the Wiener solution. If $g = 1$, the response of the beamformer is often termed the minimum variance distortionless response (MVDR) beamformer. This beamformer is attributed to Capon [7], who did the early work that lead to its development.

3.2.4 Relationship Between the Criteria

In the optimal criteria discussed in the preceding sections, we see that the expressions for the optimal weights have the same form. Despite the different values assigned to scalar β for different techniques, they all yield the same SIR, which is given as,

$$\mathrm{SIR} = \frac{\boldsymbol{w}_{\text{opt}}^H \boldsymbol{R}_s \boldsymbol{w}_{\text{opt}}}{\boldsymbol{w}_{\text{opt}}^H \boldsymbol{R}_u \boldsymbol{w}_{\text{opt}}} = \frac{\beta^2 E\{d^2(t)\} \boldsymbol{v}^H \boldsymbol{R}_u^{-1} \boldsymbol{v} \boldsymbol{v}^H \boldsymbol{R}_u^{-1} \boldsymbol{v}}{\beta^2 \boldsymbol{v}^H \boldsymbol{R}_u^{-1} \boldsymbol{v}} = E\{d^2(t)\} \boldsymbol{v}^H \boldsymbol{R}_u^{-1} \boldsymbol{v} \tag{3.30}$$

which is independent of β. The scalar β represents the scalar gain. In the minimum MSE criterion, β is given as

$$\beta_{\mathrm{MSE}} = \frac{E\{d^2(t)\}}{1+\mathrm{SIR}} \tag{3.31}$$

In the maximum (SIR) criterion, β is given as

$$\beta_{\mathrm{SIR}} = \frac{E\{d^2(t)\}}{\mathrm{SIR}} \boldsymbol{v}^H \boldsymbol{w}_{\mathrm{opt}} \tag{3.32}$$

In the minimum variance criterion, since $g = \boldsymbol{w}_{\mathrm{opt}}^H \boldsymbol{v}$, β is given as

$$\beta_{\mathrm{MV}} = g\frac{E\{d^2(t)\}}{\mathrm{SIR}} = \frac{E\{d^2(t)\}}{\mathrm{SIR}} \boldsymbol{w}_{\mathrm{opt}}^H \boldsymbol{v} \tag{3.33}$$

which is indeed identical to β_{SIR}. Furthermore, β_{MSE} can be expressed in terms of β_{MV}:

$$\beta_{\mathrm{MSE}} = \frac{E\{d^2(t)\}\beta_{\mathrm{MV}}}{gE\{d^2(t)\} + \beta_{\mathrm{MV}}} \tag{3.34}$$

The fact that the optimum weights using different criteria are all given by the Wiener solution demonstrates the fundamental importance of the Wiener-Hopf equation in establishing the theoretical adaptive beamforming steady-state performance limits.

3.3 ADAPTIVE ALGORITHMS

In the preceding section, we have shown that the optimum criteria are closely related to each other. Therefore, the choice of a particular criterion is not critically important in terms of performance. On the other hand, the choice of adaptive algorithms for deriving the adaptive weights is highly important in that it determines both the speed of convergence and hardware complexity required to implement the algorithm. In this section, we will discuss a number of common adaptive techniques. It should be pointed out that the criteria in the preceding section are described in a general sense, even though continuous time is used in each description. Obviously, they are also applicable to discrete-time signals. In this section, since we are dealing with digital adaptive beamforming algorithms, we will emphasize the digital signal processing aspects by expressing the algorithms in terms of discrete time.

3.3.1 Least Mean Squares Algorithm

The most common adaptive algorithm for continuous adaptation is the least-mean-squares (LMS) algorithm. It has been well studied and is well understood [15]. It is based on the steepest-descent method, a well-known optimization method [16], that recursively computes and updates the weight vector. It is intuitively reasonable that successive corrections to the weight vector in the direction of the negative of the gradient vector should eventually lead to the MSE, at which point the weight vector assumes its optimum value. According to the method of steepest decent, the updated value of the weight vector at time $n+1$ is computed by using the simple recursive relation

$$\boldsymbol{w}(n+1) = \boldsymbol{w}(n) + \frac{1}{2}\mu[-\boldsymbol{\nabla}(E\{\epsilon^2(n)\})] \tag{3.35}$$

It follows from (3.11) that

$$\boldsymbol{w}(n+1) = \boldsymbol{w}(n) + \mu[\boldsymbol{r} - \boldsymbol{R}\boldsymbol{w}(n)] \tag{3.36}$$

In reality, an exact measurement of the gradient vector is not possible, since this would require a prior knowledge of both $\boldsymbol{R}$ and $\boldsymbol{r}$. The most obvious strategy is to use their instantaneous estimates, which are defined, respectively, as

$$\hat{\boldsymbol{R}}(n) = \boldsymbol{x}(n)\boldsymbol{x}^H(n) \tag{3.37}$$

and

$$\hat{\boldsymbol{r}}(n) = d^*(n)\boldsymbol{x}(n) \tag{3.38}$$

The weights can then be updated as

$$\begin{aligned} \hat{\boldsymbol{w}}(n+1) &= \hat{\boldsymbol{w}}(n) + \mu\boldsymbol{x}(n)[d^*(n) - \boldsymbol{x}^H(n)\hat{\boldsymbol{w}}(n)] \\ &= \hat{\boldsymbol{w}}(n) + \mu\boldsymbol{x}(n)\epsilon^*(n) \end{aligned} \tag{3.39}$$

The gain constant μ controls convergence characteristics of the random vector sequence $\boldsymbol{w}(n)$. Note that this is a continuously adaptive approach, where the weights are updated as the data are sampled such that the resulting weight vector sequence converges to the optimum solution. Continuous adaptation works well when statistics related to the signal environment are stationary. Figure 3.3 shows the signal-flow graph representation of the LMS algorithm. The primary virtue of the LMS algorithm is its simplicity. Its performance is acceptable in many applications. However, its convergence characteristics depend on the eigenstructure of $\hat{\boldsymbol{R}}$. When the eigenvalues are widely spread, convergence can be slow and other adaptive algorithms with better convergence rates should be considered.

3.3.2 Direct Sample Covariance Matrix Inversion

One way to speed up the convergence rate is to employ the direct inversion of the covariance matrix $\boldsymbol{R}$ in (3.12) [8]. If the desired and interference signals are known

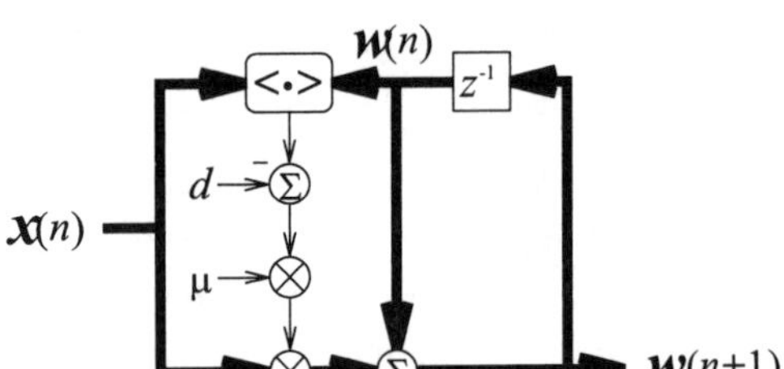

Figure 3.3 Signal-flow graph representation of the LMS algorithm.

a priori, then the covariance matrix could be evaluated and the optimal solution for the weights could be computed using (3.12), which requires direct SMI. In practice, the signals are not known and the signal environment undergoes frequent changes. Thus, the adaptive processor must continually update the weight vector to meet the new requirements imposed by the varying conditions. This need to update the weight vector without a priori information leads to the expedient of obtaining estimates of $\boldsymbol{R}$ and $\boldsymbol{r}$ in a finite observation interval and then using these estimates in (3.12) to obtain the desired weight vector. The estimates of both $\boldsymbol{R}$ and $\boldsymbol{r}$ can be evaluated, respectively, as

$$\hat{\boldsymbol{R}} = \sum_{i=N_1}^{N_2} \boldsymbol{x}(i)\boldsymbol{x}^H(i) \tag{3.40}$$

and

$$\hat{\boldsymbol{r}} = \sum_{i=N_1}^{N_2} d^*(i)\boldsymbol{x}(i) \tag{3.41}$$

where N_1 and N_2 are the lower and upper limits of the observation interval or window, respectively. Note that this is block-adaptive approach, where statistics are estimated from a temporal block of array data and used in an optimum weight equation. If a nonstationary (time-varying) environment is anticipated, block adaptation can be used, provided that the weights are recomputed periodically. It follows from (3.12) that the estimate of the weight vector is given by

$$\hat{\boldsymbol{w}} = \hat{\boldsymbol{R}}^{-1}\hat{\boldsymbol{r}} \tag{3.42}$$

If we introduce a term $\boldsymbol{e}$ to represent the errors due to the estimates, we may write

$$\boldsymbol{e} = \hat{\boldsymbol{R}}\boldsymbol{w}_{opt} - \hat{\boldsymbol{r}} \tag{3.43}$$

which can be viewed as the least squares formulation of the problem. The weight vector that is derived using the SMI method is a least squares solution.

Although the SMI approach can be shown, in theory, to converge more rapidly than the LMS algorithm, the practical difficulties that are associated with the SMI algorithm approach should be appreciated. It has two major problems: (1) increased computational complexity that cannot be easily overcome through the use of VLSI, and (2) numerical instability resulting from the use of finite-precision arithmetic and the requirement of inverting a large matrix.

3.3.3 Recursive Least Squares Algorithms

Instead of using windowing to estimate both $\boldsymbol{R}$ and $\boldsymbol{r}$, we can estimate them using the weighted sum:

$$\tilde{\boldsymbol{R}}(n) = \sum_{i=1}^{N} \gamma^{n-i} \boldsymbol{x}(i)\boldsymbol{x}^H(i) \tag{3.44}$$

and

$$\tilde{\boldsymbol{r}}(n) = \sum_{n=1}^{N} \gamma^{n-i} d^*(i)\boldsymbol{x}(i) \tag{3.45}$$

The weighting factor, $0 < \gamma \leq 1$, is intended to ensure that data in the distant past are "forgotten" in order to allow the processor to follow the statistical variations of the observable data. Factoring out the terms corresponding to $i = n$ in both (3.44) and (3.45), we have the following recursion for updating both $\tilde{\boldsymbol{R}}(n)$ and $\tilde{\boldsymbol{r}}(n)$:

$$\tilde{\boldsymbol{R}}(n) = \gamma\tilde{\boldsymbol{R}}(n-1) + \boldsymbol{x}(n)\boldsymbol{x}^H(n) \tag{3.46}$$

and

$$\tilde{\boldsymbol{r}}(n) = \gamma\tilde{\boldsymbol{r}}(n-1) + d^*(n)\boldsymbol{x}(n) \tag{3.47}$$

Using Woodbury's Identity, we obtain the following recursive equation for deriving the inverse of the covariance matrix,

$$\boldsymbol{R}^{-1}(n) = \gamma^{-1}\left[\boldsymbol{R}^{-1}(n-1) - \boldsymbol{q}(n)\boldsymbol{x}(n)\boldsymbol{R}^{-1}(n-1)\right] \tag{3.48}$$

where the gain vector $\boldsymbol{q}(n)$ is given by [15]:

$$\boldsymbol{q}(n) = \frac{\gamma^{-1}\boldsymbol{R}^{-1}(n-1)\boldsymbol{x}(n)}{1 + \gamma^{-1}\boldsymbol{x}^H(n)\boldsymbol{R}^{-1}(n-1)\boldsymbol{x}(n)} \tag{3.49}$$

To develop the recursive equation for updating the least squares estimate $\hat{\boldsymbol{w}}(n)$, we use (3.12) to express $w(n)$ as follows:

$$\begin{aligned} \hat{\boldsymbol{w}}(n) &= \boldsymbol{R}^{-1}(n)\boldsymbol{r}(n) \\ &= \gamma^{-1}\left[\boldsymbol{R}^{-1}(n-1) - \boldsymbol{q}(n)\boldsymbol{x}(n)\boldsymbol{R}^{-1}(n-1)\right] \\ &\quad \times [\gamma\boldsymbol{r}(n-1) + d^*(n)\boldsymbol{x}(n)] \end{aligned} \tag{3.50}$$

Rearranging the above equation, we can update the weight vector as follows:

$$\hat{\boldsymbol{w}}(n) = \hat{\boldsymbol{w}}(n-1) + \boldsymbol{q}(n)\left[d^*(n) - \hat{\boldsymbol{w}}^H(n-1)\boldsymbol{x}(n)\right] \tag{3.51}$$

An important feature of the recursive least squares (RLS) algorithm is that the inversion of the covariance matrix $\boldsymbol{x}(n)$ is replaced at each step by a simple scalar division. Figure 3.4 depicts a signal-flow graph representation of the RLS algorithm. The convergence rate of the RLS algorithm is typically an order of magnitude faster than that of the LMS algorithm, provided that the signal-to-noise ratio is high [15].

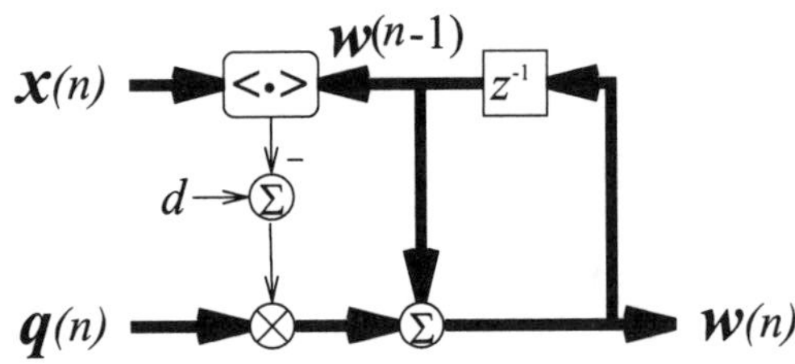

Figure 3.4 Signal-flow graph representation of the RLS algorithm.

3.3.4 Neural Networks

The resurgence of interest in the use of neural networks to solve some of the problems arising in signal-processing applications [17–19] stems from the fact that a neural network is a massively parallel interconnected network of simple elements, and their hierarchical organizations are intended to interact with the objects of the real world in the same way as a biological nervous system. More specifically, a neural network may consist of a layered or multilayered nonlinear network that tries to reproduce human intellectual capabilities through a "learning" or "adapting" process. It is an ideal tool for use in adaptive signal processing. In particular, a signal-processing problem is represented in terms of optimization where a cost function matches the energy function of a particular neural network [20]. The neural network arrives at a solution of a problem by minimizing its energy function. Recurrent networks are usually suitable for this type of processing, in which the output of a neuron is fed back as inputs to other neurons and/or to itself.

One particular type of recurrent neural network is the Hopfield network (model), which consists of a single layer of neurons that are fully connected, as shown in Figure 3.5. The output of each of the neurons is fed back to all the neurons through synaptic weights. In other words, the synaptic weights represent the feedback connectivity or conductivity. The synaptic weights can be expressed in matrix form:

$$\boldsymbol{C} = \begin{bmatrix} c_{11} & c_{12} & \cdots & c_{1K} \\ c_{21} & c_{22} & \cdots & c_{2K} \\ \vdots & \vdots & \ddots & \vdots \\ c_{K1} & c_{K2} & \cdots & c_{KK} \end{bmatrix} \tag{3.52}$$

In the Hopfield model, $c_{kl} = c_{lk}$ and $c_{kk} = 0$.

The internal potential of the kth neuron can be described by

$$\nu_k(n+1) = -\rho_k \nu_k(n) + \sum_{l=1}^{K} c_{kl} s_l(n) + b_k(n) \tag{3.53}$$

which has a temporal dynamic

$$\frac{\mathrm{d}\nu_k}{\mathrm{d}t} = -\rho_k \nu_k + \sum_{l=1}^{K} c_{kl} s_l + b_k \tag{3.54}$$

and

$$s_k = \zeta(\nu_k) \tag{3.55}$$

where ρ is called the damping coefficient, b_k is the external input signal to the neuron, and ζ is a predetermined nonlinear function. The internal potential reaches its equilibrium when $\frac{\mathrm{d}\nu_k}{\mathrm{d}t} = 0$; that is,

$$-\rho_k \nu_k + \sum_{l=1}^{K} c_{kl} s_l + b_k = 0 \tag{3.56}$$

The discrete time neuron model given in Figure 3.5 can be used for solving the above nonlinear equations if implemented using iterative methods, such as the relaxation method and the Runge-Kutta method.

The state of a Hopfield network can be described by a Lyapunov function. This is a function that becomes smaller for any change in the state of the network until a stable state is reached, when, by definition, neither the state of the network nor the Lyapunov function undergoes any further change. Hopfield demonstrated the existence of a Lyapunov function for a Hopfield network. The energy state or Lyapunov function of the network is defined as

$$\mathcal{E} = -\frac{1}{2} \sum_{k} \sum_{l \neq k} c_{kl} s_k s_l - \sum_{k} b_k s_k \tag{3.57}$$

or in matrix notation,

$$\mathcal{E} = -\frac{1}{2} \boldsymbol{s}^T \boldsymbol{C} \boldsymbol{s} - \boldsymbol{s}^T \boldsymbol{b} \tag{3.58}$$

where b_k is the external input to the kth neuron. We can gain a physical insight into this energy function by considering c_{kl} as conductance, s_k as voltage, and b_k as current. The increment in this energy function is given by

$$\Delta\mathcal{E} = -\left[\sum_{l\neq k} c_{kl}s_l + b_k\right]\Delta s_k \tag{3.59}$$

in which the state of the kth neuron is changed by Δs_k. The expression in the brackets in (3.59) drives the magnitude and sign of Δs_k, and therefore the bracketed term and Δs_k are always of the same sign. It follows that $\Delta\mathcal{E}$ is always negative or zero at a stable point. Thus, $\mathcal{E}$ fulfills the Lyapunov requirements.

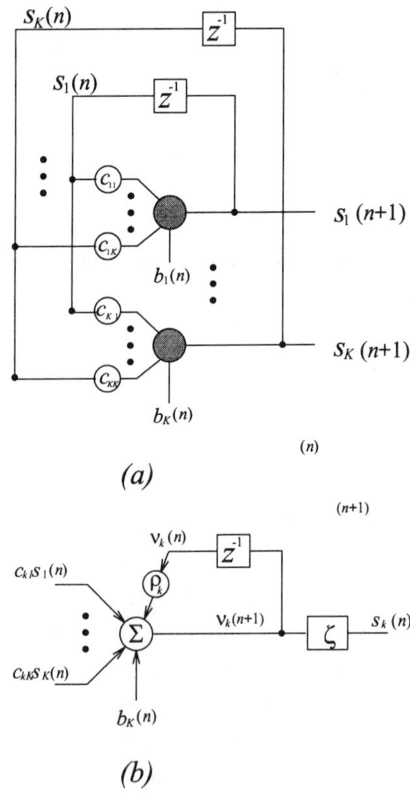

Figure 3.5 (a) Discrete-time Hopfield model of a neural network; (b) its neuron.

The fact that Lyapunov functions exist for Hopfield networks makes them suitable for solving optimization problems. In terms of energy, a stable state represents a potential well. An input vector introduced into the network represents the initial condition, which will lead to the selection of a particular well. This feature can be exploited for solving optimization problems such as minimizing the MSE in adaptive beamforming problems [21–23].

In order to map the MSE into the energy function for the Hopfield network, we carry out the following algebraic manipulations. Let us represent the beamformer output y by its real and imaginary components,

$$y = y_{\Re} + jy_{\Im} \tag{3.60}$$

where

$$y_{\Re} = \boldsymbol{w}_{\Re}^T \boldsymbol{x}_{\Re} + \boldsymbol{w}_{\Im}^T \boldsymbol{x}_{\Im} \tag{3.61}$$

$$y_{\Im} = \boldsymbol{w}_{\Re}^T \boldsymbol{x}_{\Im} - \boldsymbol{w}_{\Im}^T \boldsymbol{x}_{\Re} \tag{3.62}$$

and and the subscripts $\Re$ and $\Im$ signify the real and imaginary components, respectively. For convenience, we express both $y_{\Re}$ and $y_{\Im}$ in terms of a row vector,

$$[y_{\Re}\ y_{\Im}] = \boldsymbol{W}^T \boldsymbol{X} \tag{3.63}$$

where

$$\boldsymbol{W} = \begin{bmatrix} \boldsymbol{w}_{\Re} \\ \boldsymbol{w}_{\Im} \end{bmatrix} \tag{3.64}$$

and

$$\boldsymbol{X} = \begin{bmatrix} \boldsymbol{x}_{\Re} & \boldsymbol{x}_{\Re} \\ \boldsymbol{x}_{\Re} & -\boldsymbol{x}_{\Re} \end{bmatrix} \tag{3.65}$$

Referring back to the MSE given by (3.9), we may rewrite it as

$$\begin{aligned} E\{\epsilon^2(t)\} &= E\{(d_{\Re} - y_{\Re})^2 + (d_{\Im} - y_{\Im})^2\} \\ &= E\{d_{\Re}^2 + d_{\Im}^2 + y_{\Re}^2 + y_{\Im}^2 - 2(d_{\Re} y_{\Re} + d_{\Im} y_{\Im})\} \end{aligned} \tag{3.66}$$

It should be noted that we drop the time argument (t) for convenience. Because both $d_{\Re}^2$ and $d_{\Im}^2$ are constant, minimizing the MSE is equivalent to minimizing the following objective function:

$$\Psi = E\left\{-2\begin{bmatrix} y_{\Re} & y_{\Im} \end{bmatrix}\begin{bmatrix} d_{\Re} \\ d_{\Im} \end{bmatrix} + \begin{bmatrix} y_{\Re} & y_{\Im} \end{bmatrix}\begin{bmatrix} y_{\Re} \\ y_{\Im} \end{bmatrix}\right\} \tag{3.67}$$

Substituting (3.63) into (3.67), we obtain

$$\Psi = E\left\{-2\boldsymbol{W}^T \boldsymbol{X}\begin{bmatrix} d_{\Re} \\ d_{\Im} \end{bmatrix} + \boldsymbol{W}^T \boldsymbol{X} \boldsymbol{X}^T \boldsymbol{W}\right\} \tag{3.68}$$

If we compare (3.58) with (3.68), we find that

$$\boldsymbol{C} = -2E\left\{\boldsymbol{X}\boldsymbol{X}^T\right\} \tag{3.69}$$

$$\boldsymbol{b} = 2E\left\{\boldsymbol{X}\begin{bmatrix} d_{\Re} \\ d_{\Im} \end{bmatrix}\right\} \tag{3.70}$$

and

$$\boldsymbol{s} = \boldsymbol{W} \tag{3.71}$$

Since the exact values of $\boldsymbol{C}$ and $\boldsymbol{b}$ cannot be realized, the ensemble average can be used as an approximation instead. That is, $\boldsymbol{C}$ and $\boldsymbol{b}$ can be updated [24], respectively, by

$$\boldsymbol{C}(n) = \frac{n-1}{n}\boldsymbol{C}(n-1) - 2\boldsymbol{X}(n)\boldsymbol{X}(n)^T \tag{3.72}$$

$$\boldsymbol{b}(n) = \frac{n-1}{n}\boldsymbol{b}(n-1) + 2\boldsymbol{X}(n)\begin{bmatrix} d_{\Re} \\ d_{\Im} \end{bmatrix} \tag{3.73}$$

3.4 PARTIAL ADAPTIVITY

To elucidate the concept of partially adaptive beamforming, we need to introduce the term *degrees of freedom.* This expression refers to the number of unconstrained or "free" weights that can be used to form a beam. Adaptive beamforming can be divided into two classes: fully adaptive and partially adaptive. In fully adaptive beamforming, every available degree of freedom is used. That is, every element or beam is individually adaptively controlled to achieve maximum control of the beam pattern. For example, in the analyses in the preceding section, the number of degrees of freedom is K because there are K array elements. Although it is not explicitly stated, we assume that all K degrees of freedom are used for optimum beamforming. Fully adaptive beamforming is obviously preferred because of the following advantages:

1. The full complement of the array's degrees of freedom are available to the beamformer. Therefore, one is able to maintain the maximum control of the array's response.
2. The maximum aperture gain can be retained, since all of the array elements are used.
3. The maximum spatial resolution can be retained, since the entire array aperture is used.

When the number of array elements becomes large, however, full adaptivity may be difficult to implement in practice. Exercising a large number of an array's degrees of freedom can result in a considerable level of computational complexity, as well as a decrease in the convergence rate of the adaptive algorithm. Consequently, it is often desirable to reduce the degrees of freedom. That is, only a fraction of the available elements or beams are adaptively controlled to form the desired beam. By reducing the degrees of freedom used for adaptation, we reduce the computational complexity and improve system response time while maintaining as much control as possible over the array response. A number of strategies can be used for reducing the degrees of freedom that are presented to the adaptive algorithm, such as:

1. Judiciously selecting only a fraction of the array elements for adaptive control, thereby carrying out adaptivity at the elemental level [25];
2. Grouping the entire array into a number of subarrays, carrying out beamforming for each subarray, and adaptively controling the outputs of the subarray beamformers [26];
3. Forming beams using the entire array and adaptively controling these beams [27, 28].

Note that strategy 1 is based on element-space processing, while strategy 3 is based on beam-space processing. Strategy 2 can be considered as beam-space processing because subarray beams are processed. However, it can also be said to be an element-space technique in that the subarrays can be considered to be superelements, forming an array of equivalent aperture. That is, the adaptive control is carried out at the superelemental level. Nevertheless, these three strategies can be generalized by introducing a linear transformation $\boldsymbol{T}$, as shown in Figure 3.6.

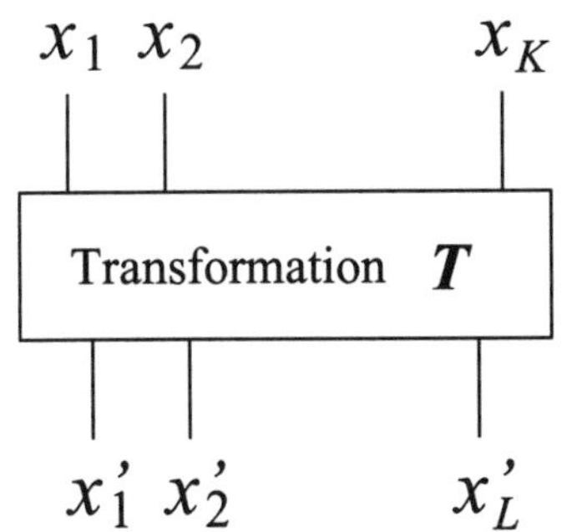

Figure 3.6 Use of linear transformation T to reduce the number of degrees of freedom from K to L.

In element-space partial-adaptivity scheme processing, only a fraction of the array elements are controlled directly. In this case, $\boldsymbol{T}$ is a $K \times L$ permutation matrix, where $L \leq K$ is the number of elements to be used. Each of the L columns in $\boldsymbol{T}$ has only one entry, which is identically equal to unity, while all other entries are equal to zero; that is,

$$\boldsymbol{T} = [\boldsymbol{t}_1, \boldsymbol{t}_2, \cdots, \boldsymbol{t}_L] \tag{3.74}$$

where

$$\boldsymbol{t}_l = [0, \cdots, 0, 1, 0, \cdots, 0]^T \tag{3.75}$$

In the case of the subarray approach, $\boldsymbol{T}$ is a $K \times L$ sparse matrix, where $L \leq K$ is the number of subarrays to be used. Each of the L columns in $\boldsymbol{T}$ has nonzero entries, which are indeed the weights for the elements of the subarray, and the rest of the entries are equal to zero, i.e.,

$$\boldsymbol{T} = [\boldsymbol{t}_1, \boldsymbol{t}_2, \cdots, \boldsymbol{t}_L] \tag{3.76}$$

where

$$\boldsymbol{t}_l = [0, \cdots, 0, t_{k+1,l}, t_{k+2,l}, \cdots, t_{k+K_s,l}, 0, \cdots, 0]^T \tag{3.77}$$

and K_s is the number of elements of a subarray.

In beam-space partial-adaptivity processing, the full array aperture is used for forming a set of orthogonal beams, a fraction of which are then weighted and combined to generate the desired adaptive response. In so doing, the maximum aperture gain can be maintained as in the case for the fully adapted array. In this case, $\boldsymbol{T}$ is a $K \times L$ matrix, where $L \leq K$ is the number of beams to be used. Each of the L columns in $\boldsymbol{T}$ has nonzero entries, which are the weights required for generating an orthogonal beam; that is,

$$\boldsymbol{T} = [\boldsymbol{t}_1, \boldsymbol{t}_2, \cdots, \boldsymbol{t}_L] \tag{3.78}$$

where

$$\boldsymbol{t}_l = [t_{1,l}, t_{2,l}, \cdots, t_{K,l}]^T \tag{3.79}$$

It follows from the transformation that the data vector for the partial adaptive approach is given as

$$\boldsymbol{x}_p = \boldsymbol{T}^H \boldsymbol{x} = [x'_1, x'_2, \cdots, x'_L] \tag{3.80}$$

Consequently, the covariance matrix is given by

$$\boldsymbol{R}_p = E\{\boldsymbol{x}_p \boldsymbol{x}_p^H\} = \boldsymbol{T}^H \boldsymbol{R} \boldsymbol{T} \tag{3.81}$$

The optimum criteria and the adaptive techniques discussed earlier can then be directly applied here.

In beam-space processing, since only a fraction of elements are now being used for adaptation, the effects of errors in the weights applied to the elements becomes of major concern when designing a partially adaptive array [25, 29]. The degree to which interference cancellation can be achieved with an element-level partially adaptive array, where errors are present in the weights, is highly sensitive to the location of the adaptive elements within the array. Consequently, for a practical design, the best choice for the location of the adaptive elements depends principally upon the error level experienced by the array weights and only secondarily on the optimum theoretical performance that can be achieved [11]. It is also found that there is an advantage in adjusting the spacing of the elements so that the controlled elements are clustered towards the center, since the adapted pattern interference cancellation tends to be independent of errors in the element weights [25, 29].

Adaptive beamforming carried out in beam space is a highly attractive solution to the problem of providing partial adaptivity to an array antenna. This is particularly true if the application inherently requires multiple beamforming, as in the case of mobile satellite communications. The use of a beam-space partially adaptive array can result in the following benefits over that of a fully adaptive array:

1. Reduction in overall cost because only a relatively small number of degrees of freedom are exercised;
2. Use of relatively simple adaptive weight constraints, thereby causing minimal degradation in the gain of the main beam and the levels of the sidelobes;
3. Reduction in the computational overload;
4. Considerably faster adaptation of the array response;
5. Compatibility with a large number of adaptive algorithms.

It should be noted, however, that there is a performance penalty associated with partially adaptive beamforming. A partially adaptive beamformer cannot converge to the same optimum solution as the fully adaptive beamformer and will experience a rapid degradation in its performance when the interference scenario calls for more degrees of freedom than available to the adaptive subsystem.

3.5 REFERENCE SIGNAL ACQUISITION

We have seen a number of criteria and algorithms for adaptive beamforming in the preceding sections. They all require some sort of reference signal in their adaptive optimization process. When people speak of reference signals, they usually mean the *a priori* and explicit information or knowledge about the signals of interest. Explicit reference can be divided into two categories: spatial reference and temporal reference. Spatial reference is mainly referred to as the angle-of-arrival (AOA) information of a desired signal. A temporal reference signal may be a pilot signal that is correlated with the desired signal, a special sequence in a packet carried by the desired signal, or a known pseudo-noise (PN) code in a CDMA system. The form of available reference depends on the particular system where adaptive beamforming is to be implemented. If an explicit reference signal is available in a system, it should be used as much as possible for less complexity, high accuracy, and fast convergence.

It is worthwhile to describe the reference signal acquisition process in a CDMA system. Adaptive beamforming is highly suitable for a CDMA system, because the spreading codes can be used as references for beamforming. The common implementation of adaptive beamforming in a CDMA wireless communications system is the use of Compton's reference generation loop [30]. A generic implementation configuration for adaptive beamforming in a CDMA wireless communications system is shown in Figure 3.7. In this configuration, the demodulation is carried out after beamforming. That is, the array output is first mixed with a CDMA-coded local oscillator signal, which is filtered and limited. The limited signal is re-modulated by mixing with the corresponding CDMA coded local oscillator signal. The remodulated signal is used as the reference signal, which is compared with the delayed array output to produce an error signal. The error signal drives the adaptive processor to update the beamforming weights. The feedback loop is a nonlinear one due to the limiter operation.

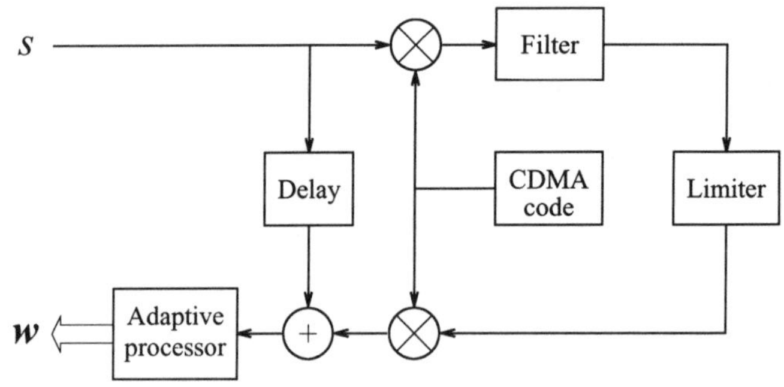

Figure 3.7 A generic configuration for adaptive beamforming in a CDMA wireless communications system.

References

[1] P. W. Howells, "Intermediate frequency sidelobe canceller," Technical report, U.S. Patent 3202990, May 1959.

[2] S. Applebaum, "Adaptive arrays," Technical Report SPL TR-66-001, Syracuse Univ. Res. Corp. Report, 1965.

[3] B. Widrow and M. E. Hoff, "Adaptive switch circuits," in *IRE WESCOM, Conv. Rec., Part 4*, pp. 96–104, 1960.

[4] L. J. Griffiths, "A simple adaptive algorithm for real-time processing in antenna arrays," *Proc. IEEE*, vol. 57, pp. 64–78, Oct. 1969.

[5] O. L. Frost III, "An algorithm for linearly constrained adaptive array processing," *Proc. IEEE*, vol. 60, pp. 926–935, Aug. 1972.

[6] L. W. Brook and I. S. Reed, "Equivalence of the likelihood ratio processor, the maximum signal-to-noise ratio filter, and the Wiener filter," *IEEE Trans. Aerosp. Electron. Syst.*, vol. 8, pp. 690–692, 1972.

[7] J. Capon, "High-resolution frequency-wavenumber spectrum analysis," *Proc. IEEE*, vol. 57, pp. 1408–1418, Aug. 1969.

[8] I. S. Reed, J. D. Mallett, and L. E. Brennen, "Rapid convergence rate in adaptive arrays," *IEEE Trans. Aerosp. Electron. Syst.*, vol. AES-10, pp. 853–863, Nov. 1974.

[9] R. T. Compton, Jr., *Adaptive Antennas: Concepts and Applications*, Prentice Hall, Englewood Cliffs, NJ, 1988.

[10] J. E. Hudson, *Adaptive Array Principles*, Peregrinus, London, 1981.

[11] R. A. Monzingo and T. W. Miller, *Introduction to Adaptive Arrays*, John Wiley & Sons, New York, 1980.

[12] W. F. Gabriel, "Adaptive arrays — an introduction," *Proc. IEEE*, vol. 64, pp. 239–272, Feb. 1976.

[13] B. D. Van Veen and K. M. Buckley, "Beamforming: A versatile approach to spatial filtering," *IEEE ASSP Magazine*, vol. 5, pp. 4–24, Apr. 1988.

[14] B. Widrow, P. E. Mantey, L. J. Griffiths, and B. B. Goode, "Adaptive antenna systems," *Proc. IEEE*, vol. 55, pp. 2143–2159, Dec. 1967.

[15] S. Haykin, *Adaptive Filter Theory*, Prentice Hall, Englewood Cliffs, NJ, 1991.

[16] W. Murray, ed., *Numerical Methods for Unconstrained Optimization*, Academic Press, New York, 1972.

[17] R. P. Lippman, "An introduction to neural net," *IEEE ASSP Magazine*, vol. 4, pp. 4–22, Apr. 1987.

[18] A. Cichocki and R. Unbehauen, *Neural Networks for Optimization and Signal Processing*, John Wiley & Sons, New York, 1993.

[19] S. Haykin, *Neural Networks — A Comprehensive Foundation*, Macmillan, New York, 1994.

[20] J. Hertz, A. Krogh, and R. G. Palmer, *Introduction to the Theory of Neural Computation*, Addison-Wesley, Redwood City, CA, 1991.

[21] S. L. Speidel, "Neural adaptive sensory processing for undersea sonar," *IEEE J. of*

Oceanic Engineering, vol. 17, Oct. 1992.

[22] B. Quach, H. Leung, T. Lo, and J. Litva, "Hopfield network approach to beamforming in spread spectrum communications," in *IEEE Proc. Seventh SP Workshop on Statistical Signal & Array Processing*, pp. 409–412, June 1994.

[23] A. Sandhu, T. Lo, H. Leung, and J. Litva, "A Hopfield neurobeamformer for spread spectrum communications," in *Sixth IEEE Int. Symposium on Personal, Indoor and Mobile Radio Communications*, Toronto, Sept. 1995.

[24] H. Leung, B. Quach, T. Lo, and J. Litva, "A chaotic neural beamformer for wireless communications," in *Proc. of PACRIM'95*, 1995.

[25] D. R. Morgan, "Partially adaptive array technique," *IEEE Trans. Antennas Propagat.*, vol. AP-26, pp. 823–833, Nov. 1978.

[26] D. J. Chapman, "Partial adaptivity for the large array," *IEEE Trans. Antennas Propagat.*, vol. AP-24, pp. 685–696, Sept. 1976.

[27] W. F. Gabriel, "Using spectral estimation techniques in adaptive processing antenna systems," Technical Report 8920, NRL, Washington, D.C., Oct. 1985.

[28] E. Brookner and J. M. Howell, "Adaptive-adaptive array processing," *Proc. IEEE*, vol. 74, pp. 602–604, Apr. 1986.

[29] R. Nitzberg, "OTH radar aurora clutter rejection when adapting a fraction of the array elements," in *EASCON*, Washington, D.C., Sept. 1976.

[30] R. T. Compton, R. J. Huff, W. G. Swarner, and A. A. Ksienski, "Adaptive arrays for communication systems: An overview of research at the Ohio State University," *IEEE Trans. Antennas Propagat.*, vol. AP-24, pp. 599–607, 1976.

Chapter 4

Error Effects on Digital Beamforming

4.1 ERROR SOURCES IN DBF ANTENNA ARRAYS

4.1.1 Random Errors

An ideal array aperture is characterized by a surface having the exact shape prescribed by theory. In an ideal DBF array, each elementary signal receives the exact weight prescribed by the designer.

No real antennas are free from physical or electrical aberrations, however. Real apertures differ from ideal apertures for many reasons. At the array level, one can identify the following sources of random error:

1. Random antenna element misplacement;
2. Element rotation about an axis parallel to the array normal;
3. Misalignment of the element axis with respect to the array normal;
4. Nonrigid antenna array structure;
5. Mutual coupling;
6. Amplitude and phase mismatch between channels;
7. (Narrowband) filter center frequency mismatch;
8. Quantization process.

Some of these sources introduce amplitude errors, others phase errors, and some both. For example, severe amplitude errors result from element rotation about an axis parallel to the array normal, which alters the polarization of the element and thereby causes a polarization mismatch with respect to the incoming radiation. Misalignment of the element axis with respect to the array normal offsets the element pattern and thereby changes its gain to an incoming signal. Phase errors are introduced by errors in the position of antenna elements. If the antenna structure is nonrigid, the face of the array may be distorted by some unexpected external force, which in turn leads to phase errors. Furthermore, the phase errors may be time-varying if the distortion is a dynamic process. Fixed but random phase errors are also introduced by transmission line length errors in the feed system.

Amplitude and phase errors are also introduced in the quantization process, since the bit length of the digital signal must be finite. Sampling can be done at baseband using two ADCs for the I and Q channels or sampling can be done directly at an IF with a single ADC with the I and Q channels generated digitally. Mutual coupling is the dominant phenomenon affecting both amplitude and phase.

Irrespective of which phenomenon is responsible for an error, its effects are on both the amplitudes and the phases of the signals across the array. Hence, any error source may be typified by errors Δa_i and $\Delta\phi_i$ to the amplitude and phase of the ith element. The purpose of error analysis is to demonstrate how such errors influence the important properties of the beam pattern of an array and to permit specifications to be established. The properties of primary concern are the gain, width, shape, and pointing direction of the main beam, as well as the sidelobe level.

4.1.2 Nonlinearities in Receivers

A typical DBF receiver is usually composed of [1]:

1. An input RF bandpass filter to eliminate image band interference from the receiver;
2. A low-noise amplifier (LNA) to establish the system noise figure;
3. One or more down-conversion stages to translate the received signal to a suitable frequency band;
4. A filter to limit the spectrum width of the signals that are quantized;
5. An ADC subsystem.

At the individual receiver level, one can distinguish the following sources of nonlinearity or distortion in active components:

1. Intermodulation products;
2. Gain compression;
3. Internally generated noise;
4. Spuriousness in mixers;
5. Amplitude and phase ripple across the signal bandwidth;
6. Gain and phase sensitivity of amplifiers to power supply voltages;
7. Uncorrelated dc offsets in the I and Q components (baseband sampling);
8. I and Q gain mismatch and phase errors (baseband sampling);
9. Analog-to-digital conversion:

    a. Quantization noise;

    b. Saturation effects;

c. Third-order nonlinearities;

d. Sampling jitter;

e. Spectral aliasing.

Intermodulation products are caused by nonlinear transfer characteristics in amplifiers and especially in mixers that rely on nonlinear devices for down-converting signals to an IF or baseband. Of special importance are the third-order intermodulation products, since they fall within the signal passband and cannot be removed by filtering, thus causing distortion.

Gain compression occurs when the output of a component can no longer linearly follow its input. At that point, the output ceases to be a scaled replica of the input and distortion is introduced.

Internally generated noise limits the minimum signal level that can be detected. It should be noted that noise is also present in passive devices such as attenuators, antenna elements.

Spurious responses in a mixer occur when harmonics of the local oscillator (LO) cause the frequency translation of unwanted frequencies at the mixer RF input to the IF. Usually the gain of amplifiers and mixers is not perfectly constant across the signal bandwidth. This depends on the bandwidth of the signal. The larger the bandwidth, the more likely the gain will vary. The same goes for the phase, which should ideally have a linear slope across the signal bandwidth. Amplifiers are gain- and phase-sensitive to their power supply voltages. For low-frequency power supply ripple, the gain and phase modulations produced by each amplifier stage are additive and cause a discrete spurious sideband to appear on the output signal.

In baseband sampling, dc offsets occur at the output of the I and Q mixers when the RF frequency is equal to the LO frequency but with a phase difference. The I and Q dc offsets are different, since no two mixers are perfectly identical. The same goes for the I and Q gain mismatch and phase errors (also with baseband sampling).

ADCs are also prone to the digital equivalent of analog nonlinearities. They also have sources of error specific to the digital domain.

Nonlinear phase characteristic in amplifiers causing amplitude modulation (AM)-to-phase modulatio (PM) conversion was not included in the list, since it is present mostly in traveling-wave-tube amplifiers (TWTA) and solid-state power amplifiers (SSPA). It can be proven that small-signal amplifiers that exhibit a limiting-type transfer characteristic do not show AM-to-PM conversion.

4.2 QUANTIZATION ERRORS IN DBF ARRAYS

4.2.1 Complex Signal Quantization Error

In digital beamforming, we deal with complex signals. That is, the complex signal $s(t)$ from each receiver consists of the in-phase component $x(t)$ and the quadrature-phase component $y(t)$; that is,

$$s(t) = x(t) + jy(t) \tag{4.1}$$

or in the polar form,

$$s(t) = r(t)e^{j\alpha(t)} \tag{4.2}$$

where

$$r(t) = \sqrt{x^2(t) + y^2(t)} \tag{4.3}$$

and

$$\alpha(t) = \arctan \frac{y(t)}{x(t)} \tag{4.4}$$

To simplify the notation, the time argument (t) will be dropped from the variables, but they should be understood as functions of time.

In the quantization process, both the in-phase and quadrature-phase components are converted into digital signals. It is common to represent the quantization effect that is due to finite word length by additive random noise [2]. That is, the quantized signal can be modeled as

$$s_q = s + n \tag{4.5}$$

where $n = u + jv$ represent the quantization noise. It is well known that n can be characterized with uniform distribution [2]. To carry out the analysis further, we assume the signal s follows the complex Gaussian distribution. This assumption is applicable to most types of communications signals. The quantized signal s_q can be written in the polar form,

$$s_q = r_q e^{j\alpha_q} \tag{4.6}$$

where

$$r_q = \sqrt{(x+u)^2 + (y+v)^2} \tag{4.7}$$

and

$$\alpha_q = \arctan \frac{y+v}{x+u} \tag{4.8}$$

Thus, the quantization error in terms of amplitude and phase are given by, respectively,

$$\Delta r = r - r_q \tag{4.9}$$

and

$$\Delta\alpha = \alpha - \alpha_q \tag{4.10}$$

In beamforming, the beam pattern is affected mostly by the phase errors of the array elements, which will be shown in a later section. Thus, we are mostly interested in the characteristics of the phase error $\Delta\alpha$.

To investigate the characteristics of $\Delta\alpha$, we start with the probability density function of $x+u$, which is given by

$$p(x+u) = \frac{1}{\sqrt{2\pi}\sigma}\int_{\frac{-\Delta q}{2}}^{\frac{\Delta q}{2}} e^{\frac{-(x+u-z)^2}{2\sigma^2}}\frac{1}{\Delta q}dz = \frac{1}{\Delta q}[\Phi(x+u+\Delta q)-\Phi(x+u-\Delta q)] \tag{4.11}$$

where Δq is the quantization step size and Φ is the normal cumulative density function (cdf), which is defined as

$$\Phi(b) = \frac{2}{\sqrt{\pi}}\int_0^b e^{-z^2}dz \tag{4.12}$$

Based on the mean value theorem, the term in the square brackets can be expressed as

$$\Phi(x+u+\Delta q) - \Phi(x+u-\Delta q) = \Delta q\varphi(\beta) \tag{4.13}$$

where $\varphi(\beta)$ is the Gaussian function; that is,

$$\varphi(\beta) = \frac{e^{-\frac{\beta^2}{2\sigma^2}}}{\sqrt{2\pi}\sigma} \tag{4.14}$$

and

$$x+u-\Delta q \leq \beta \leq x+u+\Delta q \tag{4.15}$$

It follows that

$$p(x+u) = \frac{1}{\Delta q}[\Delta q\varphi(\beta)] = \varphi(\beta) \tag{4.16}$$

As Δq approaches zero, $p(x+u)$ approaches $\varphi(x)$; that is $p(x+u)$ is approximately a Gaussian distribution when Δq is relatively small. This characteristic can also be demonstrated using numerical simulations. In Figure 4.1, the solid curve represents the cdf of Gaussian signal x. The dashed curve represents the cdf of the same signal x, which has been contaminated with the uniformly distributed noise u to represent one-bit quantization errors. The fact that the two curves are close to each other shows that $p(x+u) \simeq p(x)$. In a similar way, $p(y+v)$ can be shown to be approximately a Gaussian distribution. Therefore, the pdf of α_q is approximately a uniform distribution over the interval $[-\pi,\ \pi]$, which is demonstrated by the comparison of cdf given in Figure 4.2, where the solid curve is the cdf of α and the dashed curve is the cdf of α_q.

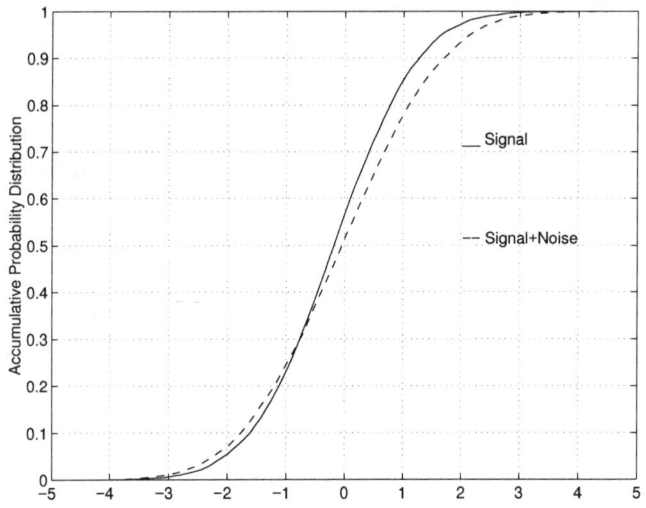

Figure 4.1 Comparison of the cdf's of a real Gaussian signal with (dashed curve) and without (solid curve) uniformly distributed real noise.

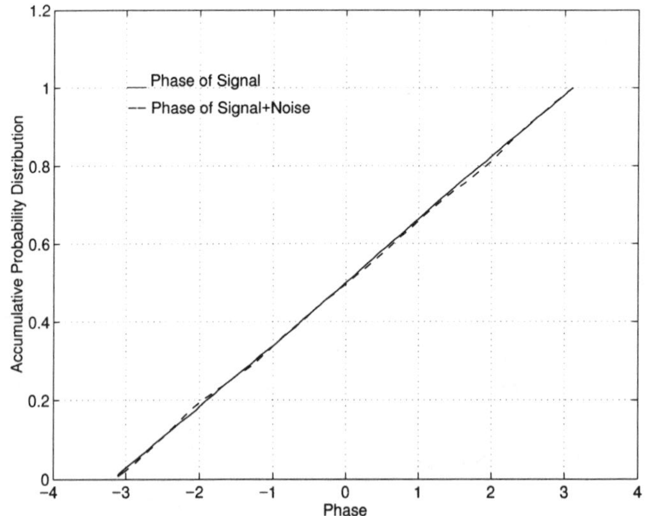

Figure 4.2 Comparison of the cdf's of phases of a complex Gaussian signal with (dashed curve) and without (solid curve) uniformly distributed complex noise.

Although α and α_q are uniformly distributed over the interval $[-\pi, \pi]$, the distribution of $\Delta\alpha = \alpha_q - \alpha$ is not easy to determine analytically. If α_q and α are mutually independent, the probability density function (pdf) of $\Delta\alpha$ will be the convolution between the pdf's of α_q and α, which is a triangle shape over the interval $[-2\pi, 2\pi]$. However, since x is correlated with $(x+u)$ and y with $(y+v)$, α_q and α are not mutually independent. That is, their correlation coefficient ρ_α, given as

$$\rho_\alpha = \frac{E\{\alpha(t)\alpha_q(t)\}}{\sqrt{E\{\alpha^2(t)\}E\{\alpha_q^2(t)\}}} \tag{4.17}$$

has some finite values. The magnitude of ρ_α depends on the number of bits, as shown in Figure 4.3. The dependence between α_q and α makes it very difficult to carry out a tractable analysis on the distribution of $\Delta\alpha$. To evaluate the distribution of $\Delta\alpha$, we have to resort to a Monte Carlo simulation. Figure 4.4 shows the distributions of $\Delta\alpha$ for one-, two-, and three-bit equalizations, respectively. In this figure, for comparison purpose, we also plot the distribution of $\Delta\alpha$ assuming α_q and α are mutually independent. A trend can be observed here. That is, as the number of quantization bits increases, the distribution of $\Delta\alpha$ will eventually become a *delta* function.

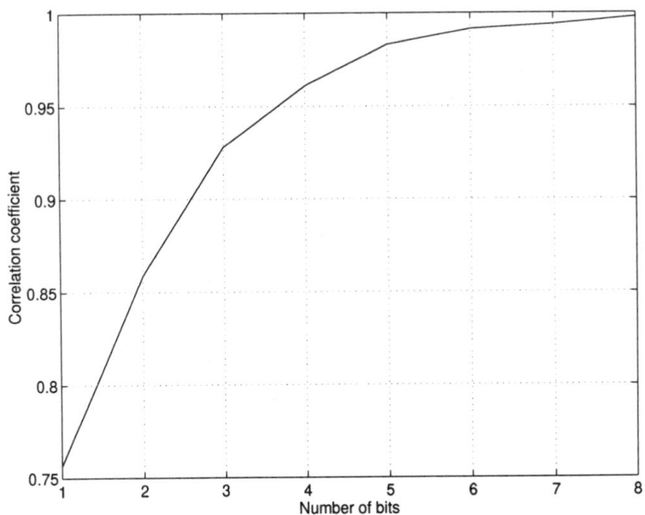

Figure 4.3 Correlation coefficient ρ_α as a function of number of bits.

The dependence between α_q and α is fortunately desirable in that the higher degree of correlation there is between α_q and α, the smaller $\Delta\alpha$ is. That is, as

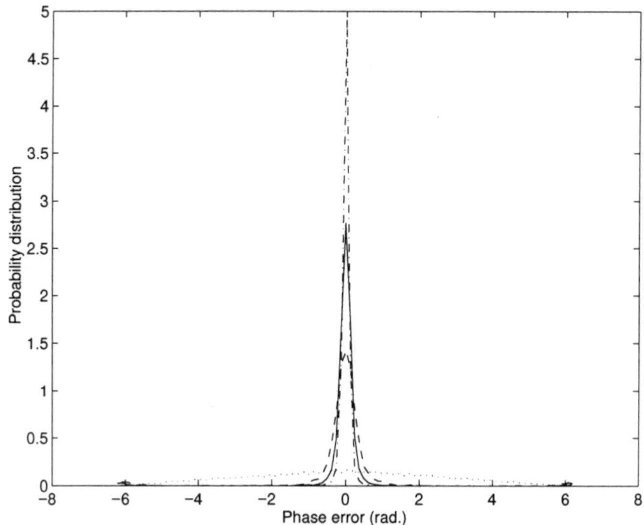

Figure 4.4 The distributions of $\Delta\alpha$ for one-bit quantization (dashed curve), two-bit quantization (solid curve), three-bit quantization (dash-dotted curve), and mutual independence between α_q and α (dotted curve).

the number of quantization bits increases, $\Delta\alpha$ approaches zero, which in turn leads to $\alpha_q = \alpha$. This is shown in Figure 4.5, where the mean square error $E\{\Delta\alpha^2\}$ is plotted as a function of number of bits.

It is interesting to compare the phase errors associated with the process of quantizing a complex signal and the errors associated with the process of digitally shifting the phase in a phased array. The effect of digital phase shifting is equivalent to that of quantizing the phase of a complex signal. That is, the phase errors associated with a digital phase shifter approximately follow a uniform distribution. Figure 4.6 compares the phase-error distributions associated with the two processes. It can be observed from the shapes of the distribution functions that the variance of phase errors associated with quantizing a complex signal using only one bit for both in-phase and quadrature-phase components is even smaller than that with phase shifting using a three-bit digital phase shifter.

4.2.2 Quantization Noise in Beamforming

The digital beamformer is required to perform the vector inner product to form a beam. Therefore, the effect of quantization noise on the beamformer output is of interest. In order to analyze this effect, it is important to understand how the

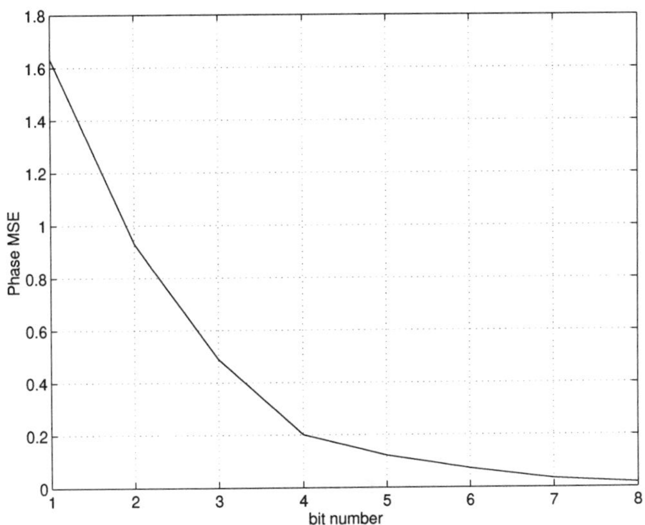

Figure 4.5 The mean square error $E\{\Delta\alpha^2\}$ as a function of number of bits.

quantization noise propagates through the beamformer. In general, the beamformer consists of a number of scalar complex multipliers and an adder.

The complex multiplier, as modeled in Figure 4.7 multiplies the signal x with the weight w, both being complex. Because both x and w are quantized, their corresponding quantization noise terms are denoted as n_x and n_w, respectively. The output of the multiplier is then given by

$$y_q = (x + n_x)^*(w + n_w) = y + n \tag{4.18}$$

where $y = x^*w$ and $n = n_x^*w + x^*n_w + n_x^*n_w$. It can be assumed that x, w, n_x, and n_w are mutually independent. By defining the signal-to-quantization-noise ratio (SQNR) as $\frac{E[|y|^2]}{E[|n|^2]}$, where

$$E[|y|^2] = \sigma_x^2\sigma_w^2 \tag{4.19}$$

and

$$E[|n|^2] = \sigma_x^2\sigma_{n_w}^2 + \sigma_w^2\sigma_{n_x}^2 + \sigma_{n_x}^2\sigma_{n_w}^2 \tag{4.20}$$

it follows that

$$\text{SQNR} = -10\lg(\frac{\sigma_{n_x}^2}{\sigma_x^2} + \frac{\sigma_{nw}^2}{\sigma_w^2} + \frac{\sigma_{n_x}^2\sigma_{n_w}^2}{\sigma_x^2\sigma_w 2}) \ \ ((\text{dB}) \tag{4.21}$$

If the real and imaginary parts of the signal x have the same dynamic range R_x and are digitized with the same number of bits b_x, the quantization noise power $\sigma_{n_x}^2$ is

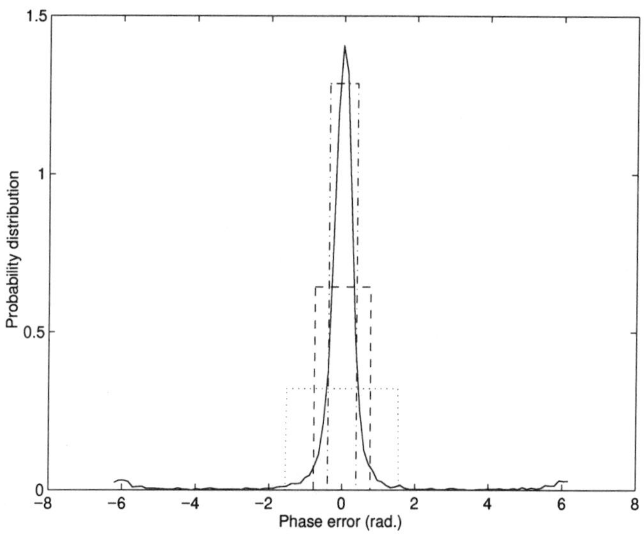

Figure 4.6 Comparison of phase-error distributions: solid curve: quantization of a complex signal using only one bit for both in-phase and quadrature-phase components ($E\{\Delta\alpha^2\} = 0.066$); dotted curve: one-bit phase shifter ($E\{\Delta\alpha^2\} = 0.82$); dashed curve: two-bit phase shifter ($E\{\Delta\alpha^2\} = 0.21$); dash-dotted curve: three-bit phase shifter ($E\{\Delta\alpha^2\} = 0.091$).

then given as

$$\sigma_{n_x}^2 = 2\frac{R_x^2}{12 * 2^{2b_x}} \tag{4.22}$$

Similarly, the quantization noise power for the weight w is given as

$$\sigma_{n_w}^2 = 2\frac{R_w^2}{12 * 2^{2b_w}} \tag{4.23}$$

Therefore, the SQNR is given as

$$\text{SQNR} = -10\lg\left(\frac{R_x^2}{6 * 2^{2b_x}\sigma_x^2} + \frac{R_w^2}{6 * 2^{2b_w}\sigma_w^2} + \frac{R_x^2 R_w^2}{36 * 2^{2(b_w+b_x)}\sigma_x^2\sigma_w^2}\right) \tag{4.24}$$

In the case that $R_x = R_w$, $\sigma_x^2 = \sigma_w^2$, and $b_x = b_w$, the SQNR becomes

$$\begin{aligned} \text{SQNR} &= 10\lg(6 * 2^{2b_x}) - 10\lg(R_x^2/\sigma_x^2) - 10\lg(2 + \tfrac{R_x^2}{6*2^{2b_x}\sigma_x^2}) \\ &= 10.8 + 6.02b_x - 10\lg(R_x^2/\sigma_x^2) - 10\lg(2 + \tfrac{R_x^2}{6*2^{2b_x}\sigma_x^2}) \text{ (dB)} \end{aligned} \tag{4.25}$$

which indicates that the SQNR increases by approximate 6 dB for every additional bit used for the quantization.

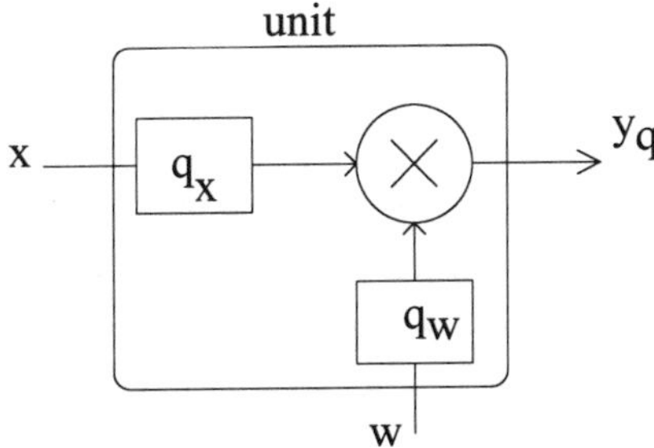

Figure 4.7 Basic unit of a beamformer.

The output of the beamformer is the sum of the outputs of all the multipliers; that is,

$$Y_q = \sum_{m=1}^{M} y_q = \sum_{m=1}^{M} (y_m + n_m) = Y + N \tag{4.26}$$

where

$$y_m = x_m^* w_m \tag{4.27}$$

and

$$n_m = n_{xm}^* w_m + x_m^* n_{wm} + n_{xm}^* n_{wm} \tag{4.28}$$

If we assume that $\{x_m^* w_m\}_{m=1}^{M}$ is a set of mutually independent variables, the average power of Y is given as

$$\sigma_Y^2 = E[|Y|^2] = M\sigma_x^2 \sigma_w^2 \tag{4.29}$$

Similarly, if $\{n_{xm}^* w_m + x_m^* n_{wm} + n_{xm}^* n_{wm}\}_{m=1}^{M}$ are mutually independent, the average power of total noise $E[|N|^2]$ is given as

$$\sigma_N^2 = E[|N|^2] = M[\sigma_x^2 \sigma_{n_w}^2 + \sigma_w^2 \sigma_{n_x}^2 + \sigma_{n_x}^2 \sigma_{n_w}^2] \tag{4.30}$$

It follows that the overall SQNR is given as

$$\text{SQNR} = 10\lg(\frac{\sigma_Y^2}{\sigma_N^2}) = -10\lg(\frac{\sigma_{n_x}^2}{\sigma_x^2} + \frac{\sigma_{n_w}^2}{\sigma_w^2} + \frac{\sigma_{n_x}^2 \sigma_{n_w}^2}{\sigma_x^2 \sigma_w^2})(\text{dB}) \tag{4.31}$$

which is identical to the SQNR for a single multiplier. That is, the overall SQNR does not depend on the number of elements.

4.2.3 Discussions

In this section, the analysis of quantization errors has been carried out in terms of the characteristics of the errors and their effects on DBF. In particular, we have determined the characteristics of the phase errors introduced by quantization and have evaluated the SQNR for the output of a digital beamformer. Based on the analysis, a number of conclusions can be derived:

1. The pdf of the phase errors, which result from quantizing the I and Q components of a complex signal, approaches the δ function as the number of quantization bits increases.
2. The phase errors due to I and Q quantization are smaller than those resulting from direct phase quantization if the same total number of bits is used.
3. In terms of digital beamforming, the overall SQNR of the beamformer output does not depend on the number of array elements. The overall SQNR increases by about 6 dB for every additional bit used for both I and Q quantization.

4.3 RANDOM ERRORS IN DBF ARRAYS

Since it is not always possible to know, in a deterministic sense, the exact nature of the errors that might be encountered in a specified antenna, the properties of the antenna must be described in statistical terms. That is, the expected value of the radiation pattern of an ensemble of antennas of similar type can be computed based on the statistics of random errors. The statistical description of the antenna properties cannot be applied to any particular antenna element, but applies to the collection of similar antennas whose errors are specified by the same statistical parameters. The analysis of the effects of random errors on a one-dimensional (1-D) array has been given in [3]. However, since two-dimensional (2-D) antenna arrays are being used more and more in various applications, there is a need to carry out the analysis of errors in 2-D arrays. Therefore, we extend the analysis used for 1-D of arrays to 2-D arrays.

4.3.1 Beam Pattern

In the error-free case, the radiation pattern of an array can be written as

$$F_0(u,v) = \frac{1}{MN}\sum_{m=1}^{M}\sum_{n=1}^{N} a_{mn}e^{(j\phi_{mn})}e^{\{j\kappa[x_m u + y_n v]\}} \tag{4.32}$$

where $a_{mn}e^{(j\phi_{mn})}$ is the complex weight of the $(m,n)^{\text{th}}$ element located at (x_m, y_n), $\kappa = 2\pi/\lambda$, $u = \sin(\theta)$, and $v = \sin(\varphi)$. θ and φ are the angles measured from the array normal in the $x-z$ plane and $y-z$ plane, respectively. If $\phi_{mn} = \kappa[x_m u + y_n v]$, $f_0(u,v)$ has the maximum response of unity.

In practice, errors must be included in the radiation pattern; that is,

$$F(u,v) = \frac{1}{MN}\sum_{m=1}^{M}\sum_{n=1}^{N}(a_{mn}+\Delta a_{mn})e^{(j\phi_{mn})}e^{(j\Delta\phi_{mn})}e^{\{j\kappa[x_m u + y_n v]\}} \tag{4.33}$$

where Δa_{mn} and $\Delta\phi_{mn}$ are the amplitude and phase errors associated with the mn^{th} element. In general, amplitude and phase errors can be assumed to be independent of each other and are each zero mean. If errors are relatively small, we can approximate the $e^{(j\Delta\phi_{mn})}$ by the first two terms of its series expansion; that is,

$$\begin{aligned}
F(u,v) &= \frac{1}{MN}\left\{\sum_{m=1}^{M}\sum_{n=1}^{N} a_{mn}e^{(j\phi_{mn})}e^{\{jk[x_n u+y_m v]\}}\right. \\
&\quad +\sum_{m=1}^{M}\sum_{n=1}^{N}\Delta a_{mn}e^{(j\phi_{mn})}e^{\{jk[x_n u+y_m v]\}} \\
&\quad +j\sum_{m=1}^{M}\sum_{n=1}^{N} a_{mn}\Delta\phi_{mn}e^{(j\phi_{mn})}e^{\{jk[x_n u+y_m v]\}} \\
&\equiv F_0(u,v)+e_a(u,v)+je_\phi(u,v)\}
\end{aligned} \tag{4.34}$$

That is, the actual pattern consists of two additional error terms: $e_a(u,v)$ due to the amplitude errors and $e_\phi(u,v)$ due to the phase errors.

The power pattern is the self-product of the radiation pattern. For the error-free radiation pattern given by (4.32), the power pattern is given by

$$F_0F_0^* = \frac{1}{M^2N^2}\sum_{m=1}^{M}\sum_{n=1}^{N}\sum_{p=1}^{M}\sum_{q=1}^{N} a_{mn}a_{pq}e^{(j\phi_{mn})}e^{[j\kappa(x_{mp}u+y_{nq}v)]} \tag{4.35}$$

and for the actual radiation pattern given by (4.33), the power pattern is given by

$$FF^* = \frac{1}{M^2N^2}\sum_{m=1}^{M}\sum_{n=1}^{N}\sum_{p=1}^{M}\sum_{q=1}^{N} a'_{mn}a'_{pq}e^{(j\Delta\phi_{mnpq})}e^{(j\phi_{mn})}e^{[j\kappa(x_{mp}u+y_{nq}v)]} \tag{4.36}$$

where

$$a'_{pq} = a_{pq}+\Delta a_{pq} \tag{4.37}$$

$$a'_{mn} = a_{mn}+\Delta a_{mn} \tag{4.38}$$

$$x_{mp} = x_m - x_p \tag{4.39}$$

$$y_{nq} = y_n - y_q \tag{4.40}$$

$$\Delta\phi_{mnpq} = \Delta\phi_{mn}-\Delta\phi_{pq} \tag{4.41}$$

The expected value of FF^* can be expressed as

$$\begin{aligned}
E[FF^*] &= \frac{A^2}{M^2N^2}\sum_{m=1}^{M}\sum_{n=1}^{N}\sum_{p=1}^{M}\sum_{q=1}^{N} a_{mn}a_{pq}e^{(j\phi_{mn})}e^{\{j\kappa(x_{mp}u+y_{nq}v)\}} \\
&\quad +\frac{1}{M^2N^2}(1-A^2)\sum_{m=1}^{M}\sum_{n=1}^{N} a_{mn}^2+\frac{1}{MN}\sigma_{\Delta a}^2
\end{aligned} \tag{4.42}$$

where $A^2 = |E[e^{j\Delta\phi}]|^2 \le 1$, and $\sigma_{\Delta a}^2$ is the variance of the amplitude error.

4.3.2 Fractional Loss in Main-Beam Gain

The error-free mainbeam gain can be evaluated by setting $\phi_{mn} = \kappa[x_m u + y_n v]$ in (4.35); that is,

$$G_0 = E\{F_0 F_0^* |_{\phi_{mn}=\kappa[x_m u + y_n v]}\} = \frac{1}{M^2 N^2} \sum_{m=1}^{M} \sum_{n=1}^{N} \sum_{p=1}^{M} \sum_{q=1}^{N} a_{mn} a_{pq} \tag{4.43}$$

When all a_{mn} are equal, $G_0 = a^2$. Likewise, the actual mainbeam gain is evaluated from (4.42):

$$G = E\{FF^*\} = A^2 a^2 + \frac{a^2}{MN}(1 - A^2) + \frac{1}{MN}\sigma_{\Delta a}^2 \tag{4.44}$$

The actual main-beam gain relative to the error-free main-beam gain is defined as the fractional loss in gain. The limiting value of the loss in gain, as the number of elements increases, is simply given by

$$\lim_{MN\to\infty} \frac{G}{G_0} = A^2 = \left| \int e^{j\Delta\phi} w(\Delta\phi) d\Delta\phi \right|^2 \tag{4.45}$$

where $w(\Delta\phi)$ is the probability density function of the phase error. That is, the loss in gain is mostly due to the phase errors. When there are many independent phenomena that contribute to the phase errors, the phase errors can be considered to follow the normal distribution. In this case, it can be shown that the fractional loss in gain is given by [3]

$$\frac{G}{G_0} = e^{-\sigma_{\Delta\phi}^2} \tag{4.46}$$

where $\sigma_{\Delta\phi}^2$ is the variance of the phase error.

4.3.3 Pointing Error

For an N-element linear array, the root-mean-square (rms) pointing error is given by the following relation [3]:

$$\sigma_u \approx \frac{\sigma_{\Delta\phi} \Delta u}{\sqrt{N}} \tag{4.47}$$

where Δu is the beamwidth of the array pattern. In the case of 2-D arrays, the total squared rms pointing error can be considered as the sum of squared rms pointing errors in both dimensions; that is,

$$\sigma_s^2 = \sigma_u^2 + \sigma_v^2 \tag{4.48}$$

It can be shown that the rms pointing error is similar to that in the 1-D case given by [3]; that is,

$$\sigma_s \propto \frac{\sigma_{\Delta\phi} \Delta s}{\sqrt{MN}} \tag{4.49}$$

where $\Delta s^2 = \Delta u^2 + \Delta v^2$. The proportionality constant is determined by the particular geometrical arrangement of element locations. Calculation of the proportionality factor requires a more rigorous analysis, which will be given as follows. The power pattern given by (4.36) can be further expressed as

$$FF^* = \frac{2}{M^2N^2} \sum_{m=1}^{M} \sum_{n=1}^{N} \sum_{p=1}^{M} \sum_{q=1}^{N} a'_{mn} a'_{pq} \cos[\kappa(x_{mp}u + y_{nq}v) + \phi_{mn} + \Delta\phi_{mnpq}] \quad (4.50)$$

In the absence of phase errors, the values of u and v that maximize FF^* are $u = u_0$ and $v = v_0$ (i.e., the main beam points at the angle (u_0, v_0)). However, with phase noise, the pointing direction will deviate from (u_0, v_0). The pointing direction can be found by setting the partial derivatives of FF^* with respect to u and v to zero; that is,

$$\begin{aligned} \frac{\partial(FF^*)}{\partial u} &= \frac{-2}{M^2N^2} \sum_{m=1}^{M} \sum_{n=1}^{N} \sum_{p=1}^{M} \sum_{q=1}^{N} a'_{mn} a'_{pq} \kappa x_{mp} \sin\left[\kappa(x_{mp}u + y_{nq}v) + \phi_{mn} + \Delta\phi_{mnpq}\right] \\ &= 0 \end{aligned} \quad (4.51)$$

and

$$\begin{aligned} \frac{\partial(FF^*)}{\partial v} &= \frac{-2}{M^2N^2} \sum_{m=1}^{M} \sum_{n=1}^{N} \sum_{p=1}^{M} \sum_{q=1}^{N} a'_{mn} a'_{pq} \kappa y_{nq} \sin\left[\kappa(x_{mp}u + y_{nq}v) + \phi_{mn} + \Delta\phi_{mnpq}\right] \\ &= 0 \end{aligned} \quad (4.52)$$

If the terms in the square brackets in the above two equations are relatively small, the sine terms can be approximated by their argument (i.e., $\sin z \approx z$), leading to

$$\kappa u = \frac{C_2B - C_1A_2}{A_1A_2 - B^2} \quad (4.53)$$

$$\kappa v = \frac{C_1B - C_2A_1}{A_1A_2 - B^2} \quad (4.54)$$

where

$$A_1 = \sum\sum\sum\sum a_{mn}a_{pq}x_{mp}^2 \quad (4.55)$$

$$A_2 = \sum\sum\sum\sum a_{mn}a_{pq}y_{nq}^2 \quad (4.56)$$

$$C_1 = \sum\sum\sum\sum x_{mp}\Delta\phi_{mnpq} \quad (4.57)$$

$$C_2 = \sum\sum\sum\sum y_{nq}\Delta\phi_{mnpq} \quad (4.58)$$

$$B = \sum\sum\sum\sum a_{mn}a_{pq}x_{mp}y_{nq} \quad (4.59)$$

The corresponding second moments are given as

$$E[(\kappa u)^2] = \frac{A_2^2E(C_1^2) + B^2E(C_2^2)}{A_1A_2 - B^2} \quad (4.60)$$

$$E[(\kappa v)^2] = \frac{A_1^2E(C_2^2) + B^2E(C_1^2)}{A_1A_2 - B^2} \quad (4.61)$$

If $\{a_{mn}\}$ are identical, B can be shown to be zero. Thus, both (4.60) and (4.61) can be simplified as

$$E[(\kappa u)^2] = \frac{\sigma^2_{\Delta\phi}}{N} \frac{1}{\sum_m x_m^2} \tag{4.62}$$

$$E[(\kappa v)^2] = \frac{\sigma^2_{\Delta\phi}}{M} \frac{1}{\sum_n y_n^2} \tag{4.63}$$

In the case of a uniformly spaced array, $x = md$ and $y = nd$, where d is the interelement spacing. The second moment of κs is then given as

$$\begin{aligned} E[(\kappa s)^2] &= E[(\kappa u)^2] + E[(\kappa v)^2] \\ &= \sigma^2_{\Delta\phi} \left(\frac{1}{N} \frac{1}{\sum_m x_m^2} + \frac{1}{M} \frac{1}{\sum_n y_n^2} \right) \\ &= \sigma^2_{\Delta\phi} \left(\frac{1}{N} \frac{1}{d_x \sum_m m^2} + \frac{1}{M} \frac{1}{d_y \sum_n n^2} \right) \\ &= \frac{6\sigma^2_{\Delta\phi}}{MN} \left(\frac{1}{d_x^2(M+1)(2M+1)} + \frac{1}{d_y^2(N+1)(2N+1)} \right) \end{aligned} \tag{4.64}$$

Therefore,

$$\sigma_s = \frac{\sqrt{6}\sigma_{\Delta\phi}}{\kappa\sqrt{MN}} \left(\frac{1}{d_x^2(M+1)(2M+1)} + \frac{1}{d_y^2(N+1)(2N+1)} \right)^{1/2} \tag{4.65}$$

If the values of both M and N are large, then

$$\frac{\lambda^2}{d_x^2(M+1)(2M+1)} \simeq \frac{\Delta u^2}{2} \tag{4.66}$$

and

$$\frac{\lambda^2}{d_y^2(N+1)(2N+1)} \simeq \frac{\Delta v^2}{2} \tag{4.67}$$

It follows that

$$\sigma_s = \frac{\sqrt{3}\sigma_{\Delta\phi}\Delta s}{2\pi\sqrt{MN}} \tag{4.68}$$

The fractional pointing error is therefore given by

$$\varepsilon_s = \frac{\sigma_s}{\Delta s} = \frac{\sqrt{3}\sigma_{\Delta\phi}}{2\pi\sqrt{MN}} \tag{4.69}$$

From the design point of view, the number of elements required to satisfy a given-error tolerance ε_s should be

$$MN = \left(\frac{\sqrt{3}}{2\pi\varepsilon_s}\right)^2 \approx \frac{0.076}{\varepsilon_s^2} \tag{4.70}$$

It shows the relationship between ε_s and MN, the minimum number of elements required to satisfy this condition for various pointing-error tolerances.

4.3.4 Sidelobes

The sidelobe level at any given angle is the sum of the value at that angle due to the design pattern plus a random quantity due to both the amplitude and phase error. That is, the term in (4.44),

$$\frac{a^2(1-A^2)+\sigma_{\Delta a}^2}{MN} \tag{4.71}$$

causes the sidelobe level to increase. Since the main lobe gain is a^2A^2, the expected value of the increase in sidelobe level, relative to the main lobe, is given by

$$\Delta_s = \frac{a^2(1-A^2)+\sigma_{\Delta a}^2}{MN * a^2A^2} = \frac{1}{MN}\left[\frac{1}{A^2}\left(1+\frac{\sigma_{\Delta a}^2}{a^2}\right)-1\right] \tag{4.72}$$

In the case of Gaussian phase errors, $A^2 = e^{-\sigma_{\Delta\phi}^2}$ and the increase in sidelobe level is

$$\Delta_s = \frac{1}{MN}\left[e^{-\sigma_{\Delta\phi}^2}\left(1+\frac{1}{\text{SNR}}\right)-1\right] \tag{4.73}$$

where $\text{SNR} = \frac{a^2}{\sigma_{\Delta a}^2}$. For example, in the case where one bit is used for quantizing both the in-phase and quadrature-phase components, $\sigma_{\Delta\phi}^2 = 0.066$ and $\text{SQNR} = \frac{1}{12}$ for $\frac{R_x^2}{\sigma_x^2} = 2$. It follows that

$$\Delta_s = \frac{12.85}{MN} \tag{4.74}$$

That is, only if the value of MN is much greater than 12.85, will the increase in sideloble level be small.

4.3.5 Effect of Element Failure

If $M'N'$ denotes the number of elements that are operative, $MN - M'N'$ represents the number of failure elements. The apparent effect of element failure is to reduce the gain in the power pattern FF^* by a factor of $(\frac{M'N'}{MN})^2$. Since the element failure is a random process, the probability $p_o = \lim_{MN\to\infty}\frac{M'N'}{MN}$ may be assigned as the operative probability of any given element. Replacing a in (4.44) with $p_o a$ gives

$$E[FF^*] = p_o^2A^2F_0F_0^* + \frac{p_o^2a^2(1-A^2)+\sigma_{\Delta a}^2}{MN} \tag{4.75}$$

That is, the expected value of the main-beam gain, relative to the error-free array, is reduced by a factor of p_o^2.

The effects in terms of pointing error are due to the number of operative elements $M'N'$. Since the standard deviation of pointing error is inversely proportional to the square root of the number of elements, the pointing error increases from σ_s to $\sigma_s(p_o)^{-1/2}$.

4.3.6 Discussions

In this section, we have carried out an analysis of the effects of random errors on a 2-D antenna array and have derived the expressions of the main-beam gain loss, pointing error, sidelobe level, and effect of element failure. Based on the analysis results, it can be observed that the effects of errors on a 2-D array are similar to those of a 1-D array. That is:

1. The main-beam gain is relatively sensitive to random errors and, therefore, most beamforming performance specifications on the magnitude of the tolerable errors (phase errors in particular) are given in terms of the allowed degradation in main-beam gain or the allowed increase in sidelobe level.
2. The loss in main-beam gain is caused mostly by phase errors if the number of array elements is relatively large.
3. Increasing the number of array elements reduces the pointing error.
4. A lower sidelobe level is likely to be maintained when the number of array elements is large.
5. The loss in gain, pointing error, and sidelobe level is independent of the direction at which the main beam points.
6. The loss in main-beam gain is proportional to the square of the probability of an element being operative.
7. The pointing error is inversely proportional to the square root of the probability of an element being operative.

4.4 NONLINEARITIES IN DBF ARRAYS

4.4.1 Modeling of Nonlinearities

Nonlinear models can be divided into two groups [4]: *memoryless* or *instantaneous models* and *models with memory.* In the memoryless model the output of the device at any instant t is a function of the input signal at the same instant only. In addition, the transfer characteristics are frequency independent. These models are very popular, since they are fairly accurate in characterizing a great variety of devices and systems.

If the output of a system or device at any instant also depends on some of the past input values, the system is said to have memory and is sometimes referred to as a *dynamic system.* Systems with frequency-dependent transfer characteristics are also considered to be systems with memory. Nonlinearities with memory are more difficult to handle. A number of general input-output-type models have been proposed for nonlinear systems with memory. However, they do not provide an efficient implementation with easily measured parameters. Thus, they are not accepted as readily as the transfer-function model is in the linear system case [4].

The memoryless narrowband signal model is suitable for microwave devices or systems with a wide bandwidth, because over any relatively small interval of the operating frequency band, the transfer characteristics are nearly independent of frequency [4]. Two principal nonlinearities present in small-signal amplifiers and mixers that will introduce distortion at the output of a receiver are intermodulation products and gain compression. Intermodulation products are caused by nonlinear transfer characteristics in amplifiers and especially in mixers that rely on nonlinear devices for down-converting signals to an IF or baseband. Of special importance are the third-order intermodulation products, since they fall within the signal passband and cannot be removed by filtering. Gain compression occurs when the output ceases to be a scaled replica of the input signal; that is, it saturates.

In modulated carrier radar receiver systems, most devices, including the nonlinear ones, are bandpass in nature. Nonlinear devices are most commonly modeled by a memoryless nonlinearity that exhibits nonlinear gain (AM/AM) and amplitude-to-phase conversion (AM/PM). These instantaneous representations are generally valid for bandpass signals that are considered sufficiently narrowband; that is, the device transfer characteristic is essentially frequency-independent over the signal bandwidth.

The output of a nonlinear system with a bandpass input includes spectral components centered around the frequencies $\pm n f_c$, $n = 0, 1, 2, \cdots$. The assumption that the memoryless system is bandpass implies that only the spectral components around $\pm f_c$ are processed by the system (the first-zone output). Therefore, the output of a nonlinearity in a bandpass system can be thought of as being followed by a zonal filter.

Consider a narrowband signal $x(t)$ with a spectrum centered at f_c:

$$x(t) = A(t) \cos\left[2\pi f_c t + \phi(t)\right] \tag{4.76}$$

where $A(t)$ = envelope modulation;
$\phi(t)$ = phase modulation.

In general, the output of a nonlinear device can be expressed as [4]

$$y(t) = f\left[A(t)\right] \cos\left\{2\pi f_c t + \phi(t) + g\left[A(t)\right]\right\} \tag{4.77}$$

where $f(\ )$ represents the amplitude nonlinearity (AM/AM) and $g(\ )$ represents the phase nonlinearity (AM/PM). Relations of the form given by (4.77) characterize what are referred to as *envelope nonlinearities.* Furthermore, in an envelope nonlinearity, the nonlinear part of the output depends only on the envelope or modulus of the input signal [5, 6].

The envelope model of a bandpass nonlinearity is a very attractive form for use in simulations, because the carrier f_c is explicit and can therefore be easily transformed to a lowpass equivalent model. The AM/AM and AM/PM envelope nonlinearity has a very simple lowpass representation. The complex envelope of the signal $x(t)$ is

$$\tilde{x}(t) = A(t)\, e^{j\phi(t)} \tag{4.78}$$

and the complex envelope or lowpass equivalent of $y(t)$ is given as

$$\tilde{y}(t) = f\,[A(t)]\; e^{j\,[\phi(t) + g\,(A(t))]} \tag{4.79}$$

which is referred to as a *serial* or *polar* model. Figure 4.8 is a block diagram representation showing explicitly how to implement the model in the complex envelope notation.

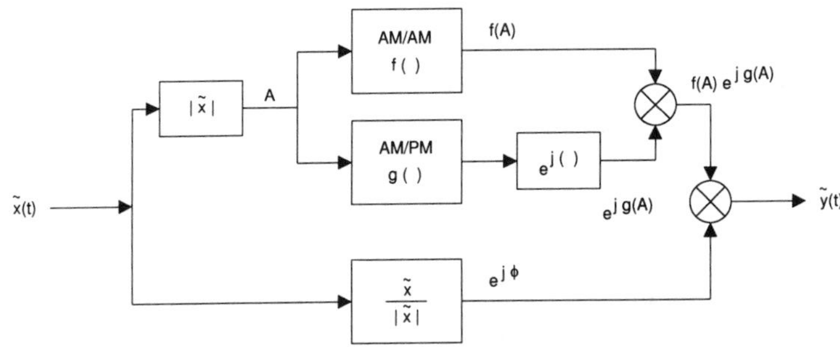

Figure 4.8 Explicit simulation model of AM/AM and AM/PM envelope (bandpass) nonlinearity. Both A and ϕ are functions of time.

Nonlinearities in Small-Signal Amplifiers

Amplifiers are one of the most important and basic elements of radar receiver systems. Practical amplifiers exhibit nonlinear behavior, which can become dominant when the input of the amplifiers becomes too high. In modeling small-signal amplifiers, four parameters can be used to describe practical amplifiers:

1. Small-signal gain (dB);
2. Noise figure (dB);
3. 1-dB compression point (dBm);
4. Third-order intercept point (dBm).

The input/output transfer characteristic of small-signal amplifiers can be characterized as being of the limiting type; that is, as the input signal level is increased at some point the output signal level will begin to saturate. Eventually, the output level will no longer increase, even if the input is increased further. The main cause of the limiting effect is due to the fixed supply or supplies of the amplifier, which will prevent the output voltage from going beyond a certain limit. All limiters are examples of memoryless nonlinearities. Desirable attributes for a limiter model are as follows [7]:

1. It has to be linear for small signals.
2. It has to be asymptotic to a definite limit level.
3. It should preferably be an analytic closed-form.
4. It should be capable of modeling knees of various sharpness, preferably by changing a single parameter.

Several limiter models have been proposed in the literature. A closed-form model with three adjustable parameters appears to offer the best compromise. Cann suggested such a parametric form for a limiter that meets the above criteria. For a baseband device, the limiter can be modeled using [4, 7]

$$y = \frac{L\,\mathrm{sgn}(x)}{\left[1 + \left(\frac{l}{\mid x \mid}\right)^{s}\right]^{\frac{1}{s}}} \tag{4.80}$$

where x = instantaneous input voltage;
y = instantaneous output voltage;
L = asymptotic output level as $\mid x \mid \rightarrow \infty$;
l = input limit level;
s = knee sharpness.

On the other hand, a bandpass limiter consists of a limiter followed by a bandpass filter, which selects only the signal existing about the carrier frequency (the first-zone output). Thus we define $\mid \tilde{x} \mid$ as the instantaneous envelope input (i.e., the magnitude of the complex envelope), and $\mid \tilde{y} \mid$ as the instantaneous output envelope. The envelope of the first-zone bandpass output is given by

$$\mid \tilde{y} \mid = f(\mid \tilde{x} \mid) \tag{4.81}$$

where the function f describes the magnitude of the output envelope as a function of the input envelope. A family of curves for f based on (4.80) is shown in Figure 4.9. This represents the situation in which a nonlinear device is followed by a bandpass filter and, for a single frequency sine wave input, only the fundamental component of the output spectrum is of interest.

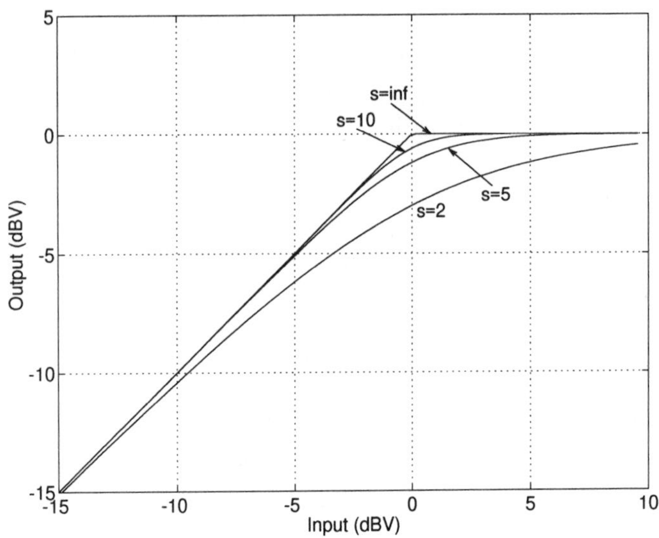

Figure 4.9 Transfer characteristics of bandpass limiter.

The limiter model does not include phase distortion. However, it can be proven through Fourier series expansion and selecting only the first-zone output from the memoryless nonlinearity that the limiter family, of which Cann's model belongs, cannot generate phase distortion (i.e., AM/PM conversion) [4]. This is also true in practice, since the maximum value of small-signal amplifiers normally never exceeds a few degrees per dB near amplifier saturation, and is generally ignored. Thus (4.79) is simplified to

$$\tilde{y}(t) = f\left[A(t)\right] e^{j\phi(t)} \tag{4.82}$$

In addition, the explicit block diagram representations of the AM/AM and AM/PM envelope nonlinearity and quadrature envelope nonlinearity reduce to the same block diagram (Figure 4.10).

Of the three parameters in (4.80), only two of them can be evaluated easily under some specific conditions. The third parameter, the knee sharpness, is not as easy to find even when the other two parameters are known. This is because (4.80) is a nonlinear equation. Hence, there is no simple equation relating s to L, l, and the 1-dB compression point. In the case of practical amplifiers, the best approach is to use curve fitting based on the measured data for the input and output levels to obtain the three parameters. Figure 4.11 shows the nonlinear model curve fitted to the measured data points of a particular amplifier. The linear gain is 8.4 dB and the 1-dB compression point is 14.7 dB.

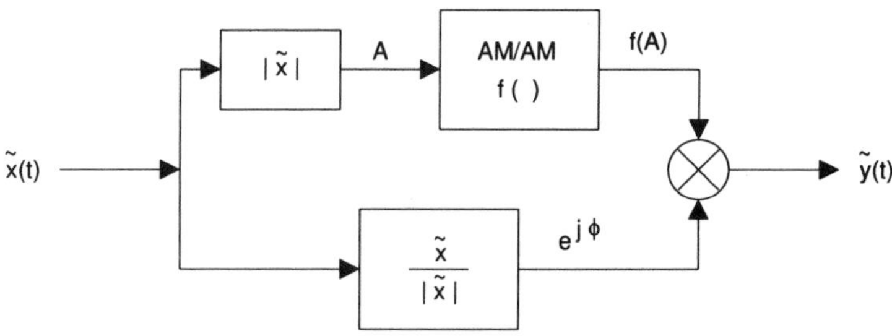

Figure 4.10 Explicit simulation model of AM/AM-only envelope nonlinearity.

The model presented is based on the idea that the response to modulated inputs can be determined, at least approximately, by using a model that is calibrated to produce correct results for single-tone inputs. This idea seems to be correct for very narrowband inputs that can be viewed as slowly varying single tones. For wider-band inputs, this assumption may not hold. In such cases, it would seem natural that a wider-band signal should be used to characterize the device. Furthermore, there is no guideline for the accuracy of the model for wideband inputs.

Modeling Third-Order Intermodulation Distortion

One problem with the model presented thus far involves the production of intermodulation components. All the models described in the literature will yield intermodulation products. However, since these models are only characterized to yield correct single-tone measurements it is not expected that the magnitudes of these intermodulation products are correct. For some, the only way these products would be close to the correct magnitude would be if somehow the model had some physical basis, which has yet to be demonstrated [8].

Through simulations of the model of the limiter with a two-tone input at different levels and also with different values of knee sharpness, it was found that besides the two fundamental tones, there are odd-order harmonics, because the output of the limiter saturates symmetrically [9], as shown in Figure 4.12. It was also observed that the slope of the third-order intermodulation products was equal to the knee sharpness plus 1 (i.e., $s + 1$) for small-signal levels. Hence, the third-order products will have a slope of 3 (on a decibel scale) if $s = 2$. In the binomial series expansion of (4.80), the exponent of the second term in the series is $s + 1$. However, the examination of the second term in the series does not reveal any information about the frequency of the intermodulation products. If the parameters of the limiter model are determined through curve fitting, it is most likely that $s \neq 2$. Thus, for small signals, the level of the third-order intermodulation products will not be

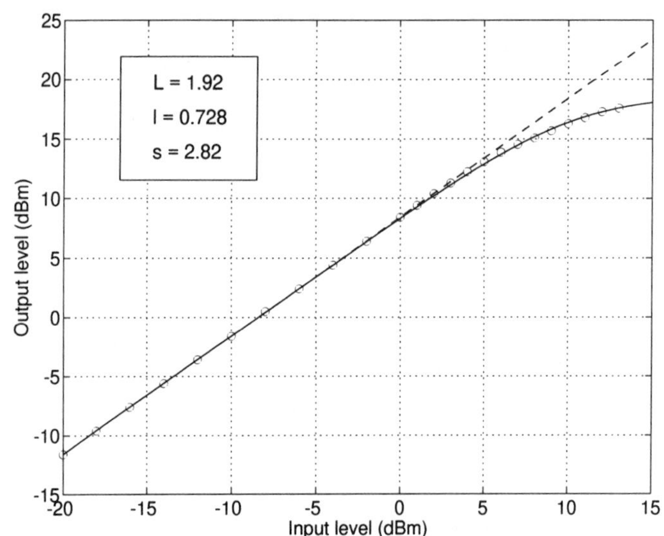

Figure 4.11 Limiter model parameters fitted to measured data (MAR-4 amplifier at 75 MHz). The circles represent the measured data, the solid line is the fitted curve, and the dashed line is the ideal linear gain.

correct in most cases.

One way to have better control over the level of the third-order intermodulation products in the model is to deliberately introduce a parameter that will generate such products. This can be accomplished by using the first and third terms in the Taylor series expansion of a nonlinear device; that is,

$$y = \alpha_1 x + \alpha_3 x^3 \tag{4.83}$$

Since the gain is already included in the limiter model, (4.83) becomes

$$y = (x + \alpha_3 x^3)\, k \tag{4.84}$$

where x = peak envelope;
k = linear gain;
α_3 = parameter to adjust the third-order intercept point, P_{IP_3}.

The relation between the third-order intercept point, P_{IP_3}, and α_3 is given by

$$P_{IP_3} = 20 \log k - 10 \log \alpha_3 - 10 \log (Z\, P_{\text{ref}}) \tag{4.85}$$

where Z = input/output impedance of the device
P_{ref} = reference power.

The introduction of the third-order intermodulation products parameter, α_3, in the model is shown schematically in Figure 4.13. This block diagram represents the function $f(\ )$ in Figure 4.10.

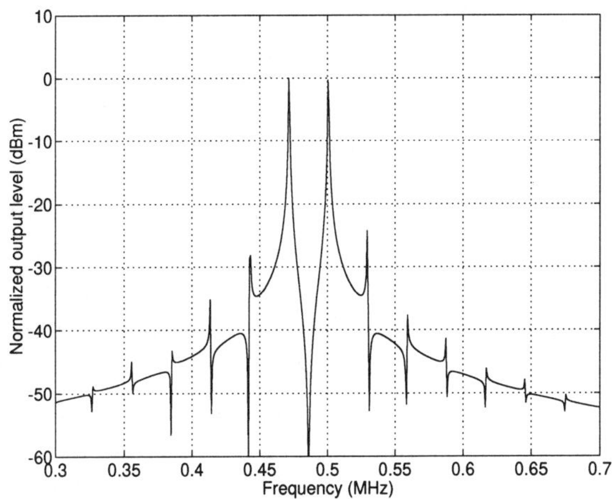

Figure 4.12 Typical output spectrum of the limiter model in the nonlinear region with a two-tone input.

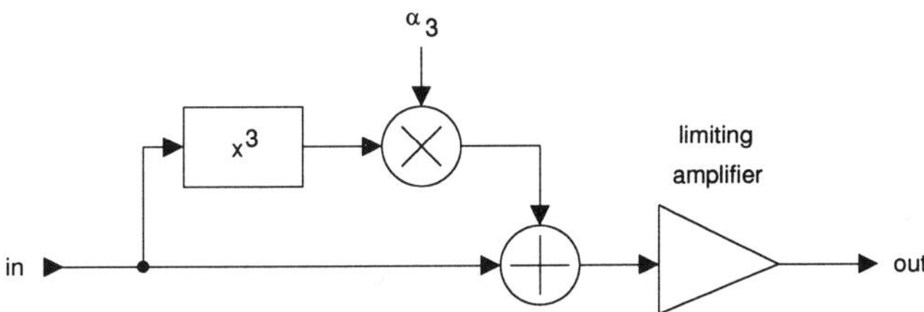

Figure 4.13 Nonlinearity including third-order intermodulation distortion and limiting amplifier.

Mixers

There are several types of mixers, such as single-ended, balanced (single, double, and triple), and image rejection. All mixers will translate both RF sidebands to an IF. That is $f_{\mathrm{IF}} = f_{\mathrm{LO}} - f_{RF_1}$ in the case of the lower sideband and $f_{\mathrm{IF}} = f_{RF_2} - f_{\mathrm{LO}}$ for the upper sideband. The sideband in which the wanted signal is present is called the real or desired frequency, and the unused sideband is called the image frequency. Diodes (usually Schottky-barrier) or field effect transisters (FET) are used for mixers. Mixers can have either *conversion loss* or *conversion gain*. Diode mixers can have conversion gain by adding small-signal amplifiers at each of their three ports.

To model the subsystem behavior of a mixer, the main modeling parameters include

1. Conversion loss or conversion gain (dB);
2. 1-dB compression point (dBm);
3. Third-order intercept point (dBm);
4. Noise figure (dB);
5. Image rejection (dB).

Since mixers exhibit gain compression and third-order intermodulation distortion just like small-signal amplifiers, the nonlinear amplifier model is used in the mixer model. The parameters of the nonlinear amplifier model can be adjusted to give a gain or a loss, which is exactly what we need. For passive mixers, the conversion loss is typically 7 dB. The 1-dB compression point is usually 3 to 8 dB lower than the local oscillator power level and the third-order intercept point is approximately 10 to 14 dB above the 1-dB compression point.

Since there is no explicit carrier frequency in the modeling, either RF, IF, or LO, the RF input spectrum of the mixer has to be translated in some way to obtain the desired results at the IF output. In the complex domain, the shifting is done by multiplying the input signal by $e^{j\omega t}$. This is done twice, once for each sideband and the two translated signals are added together to give the IF output.

Image rejection is easily included in the model by adding an attenuator in one of the translated frequency path (i.e., the image path). The attenuation is equal to the amount of image rejection. Thus, an image rejection mixer is modeled by setting the attenuation greater than zero.

Figure 4.14 is a block diagram of the mixer. The nonlinear amplifier block is the same as for the small-signal amplifier model. The IF frequency is given by

$$f_{\mathrm{IF}} = | f_{\mathrm{RF}} + f_{\mathrm{offset}} - f_{\mathrm{LO}} | \tag{4.86}$$

The offset parameter, f_{offset}, allows for a signal with a certain bandwidth not centered exactly at f_{RF} to still be translated to the center of the IF band. In the real path, the mixer input needs only to be translated by $-f_{\mathrm{offset}}$ (in the complex domain). In the image path, the mixer input is translated by $-f_{\mathrm{offset}} + 2(f_{\mathrm{RF}} + f_{\mathrm{offset}} - f_{\mathrm{LO}})$. In this way the upper sideband or the lower sideband can be selected as the real signal.

4.4.2 Receiver Nonlinearity Effects on Fixed Beamforming

The effects of receiver nonlinearities are investigated on a 32-element receiving array with a fixed beam pattern. The quiescent beam pattern is formed with 30-dB sidelobe Dolph-Chebyshev weighting. The element spacing is $\lambda/2$. Both types of errors are investigated separately.

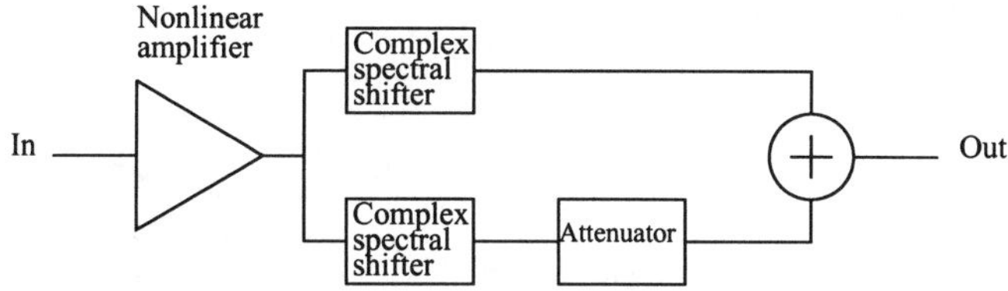

'igure 4.14 Block diagram of the mixer model.

In the simulations, a single tone was used for the signal source, and different ıterference sources were used in the signal-plus-interference cases. Two values of IR were used: -30 and 0 dB. In each case the SIR of was kept constant while the ower of the sources was varied across the dynamic range of the receivers and beyond .e., into saturation). The signal is coming from broadside while the interference rrives at 19.4° (i.e., in the direction of a sidelobe peak).

The interference source was changed for the different scenarios. The character-tics of the LNA and the mixer are given in Table 4.1. The characteristics of the ıixer were chosen so that the $P_{1\text{dB}}$ of the LNA would set the $P_{1\text{dB}}$ of the receiver lements. Figure 4.15 shows the LNA and mixer transfer characteristics plotted on he same graph. The overall gain of one receiver element is 23 dB.

Table 4.1
LNA and Mixer Characteristics.

	LNA	*Mixer*
G	+30 dB	-7 dB
$P_{1\text{dB}}$	-7 dBm	+3 dBm
P_{sat}	0 dBm	+10 dBm
P_{IP3}	+3 dBm	+13 dBm

To evaluate the results, the beam pattern formed for the ideal case (i.e., no onlinearities involved in the receiver) and the nonideal cases are compared. The eam pattern for the linear receiver case is shown with dashed lines while the eam pattern for the nonlinear receiver case is shown with solid lines. Figure 4.16 lustrates the performance of the receiving array for the desired-signal-only case for vels of (a) −60 dBm and (b) −20 dBm at the input of each receiver element. There practically no degradation in the beam pattern, except when the receivers operate the compression region as in (b), but the beam pattern no longer increases. Since ıere is only one tone, intermodulation products do not occur.

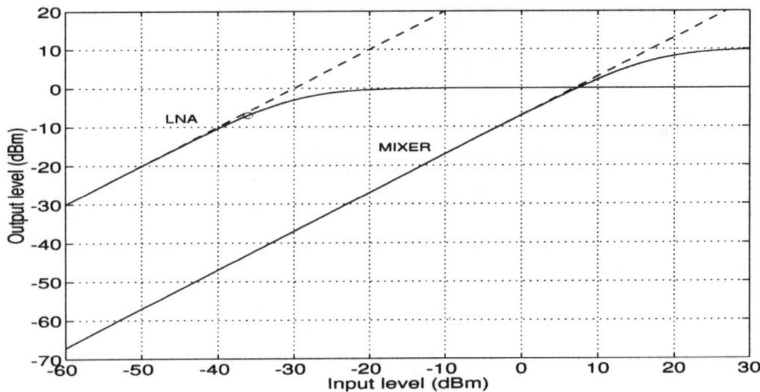

Figure 4.15 LNA and mixer transfer characteristics.

Figure 4.17 shows the results for the signal-plus-interference case with SIR = −30 dB for interference power levels of (a) −50 dBm and (b) −10 dBm at the input of each receiver element. The interference source is a single tone. For small input levels as in (a), there is little degradation in the beam pattern. For large input levels as in (b), the compression effect is the same as for the signal-only case. It can be observed that there is some degradation in the beam pattern in the appearance of a small "phantom" signal at about 42°.

Figures 4.18 to 4.20 present the results for the signal-plus-interference case with SIR = 0 dB. Both sources are single tones. The figures show the output spectrum of one receiver element in (a) and the beam pattern in (b). The power level of the sources was varied so that the input level of the receiver elements changed from −70 dBm to −10 dBm. In the case of large inputs, the intermodulation products can be clearly seen as the input level is increased. When intermodulation products appear at the output of the receivers, phantom signals also appear on both sides of the two major lobes. As the levels keep increasing, the saturation in the receivers affects the major lobes. For highly saturated receivers, multiple phantom interferers severely degrade the array output pattern.

Figures 4.21 and 4.22 show the results for the signal-plus-interference case when the interference source is wideband noise with SIR = 0 dB. The power of the sources was varied in a way similar to that of the signal-plus-tone interference case. The array output pattern degrades as the level at the input of the receivers increases but there are no definite phantom lobes appearing in the beam pattern as in the single-tone interference case.

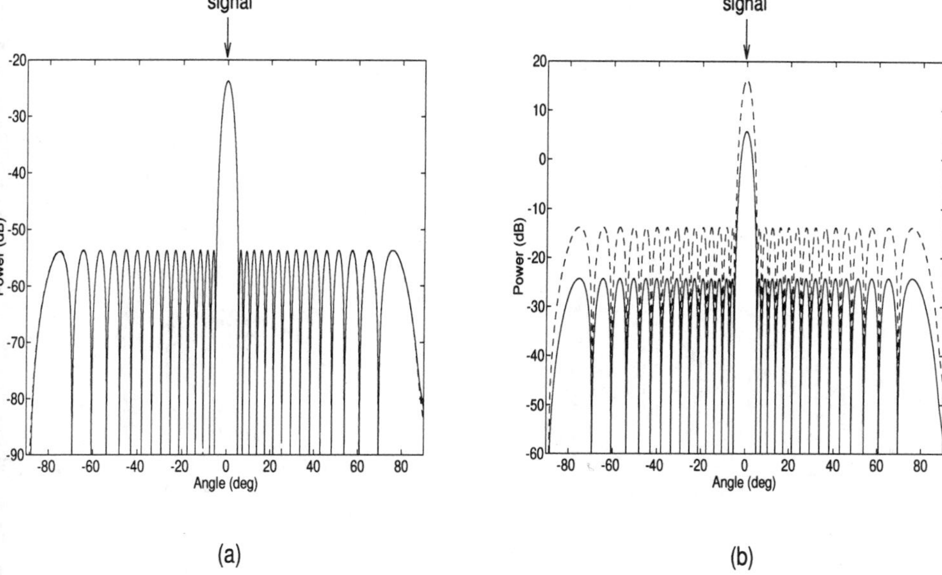

Figure 4.16 Desired-signal-only case: (a) −60 dBm; (b) −20 dBm.

4.4.3 Receiver Nonlinearity Effects on Adaptive Beamforming

To analyze the nonlinearity effects on adaptive beamforming, the same simulation setting as the one in the case of a fixed pattern is used, but without the Dolph-Chebyshev weighting. Instead, the LMS adaptive beamforming technique is applied. It is assumed that a perfect reference signal is obtainable so as to investigate the effects of noise and nonlinearities on the LMS adaptive beamforming.

Figures 4.23, 4.24, and 4.25 show the adapted beam patterns using the nonlinear receivers for different SNR levels. A number of observations can be made here. First, the phantom lobes have been greatly reduced when adaptive beamforming is applied. That is, the effects of nonlinearities can be reduced using the adaptive processing. Second, in the cases where SNR is 10, 20, and 40 dB, a null is generated towards the interferer (at 19.4°) or at the proximity. In these cases, 40-dB null with respect to the main beam is steered in the direction of the interferer. In the case where SNR is 0 dB, the LMS algorithm fails to place a null in the direction of the interference source. Finally, by comparison of the results for high and low SNR levels, it is also clear that the receiver noise affects both the main lobe and sidelobes in the adapted beam pattern. For clarity, we compare in Figure 4.26 the beam patterns of the linear receivers and the nonlinear receiver for the case of 0 dB SNR. In addition to causing a higher sidelobe level, the receiver nonlinearities disable the algorithm to steer a null towards the interferer.

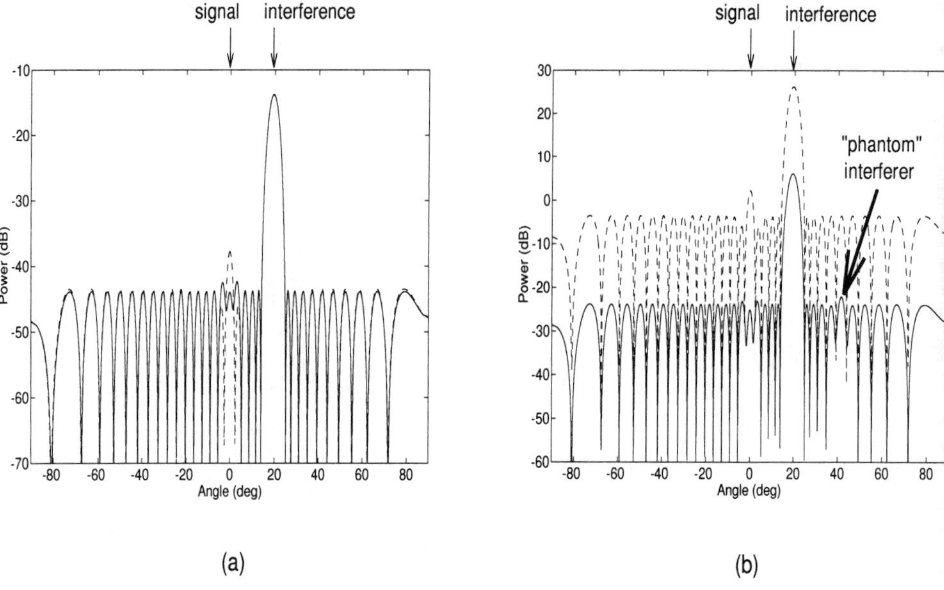

Figure 4.17 Signal-plus-interference case (single-tone interferer, SIR = -30 dB): (a) −50 dBm; (b) −10 dBm.

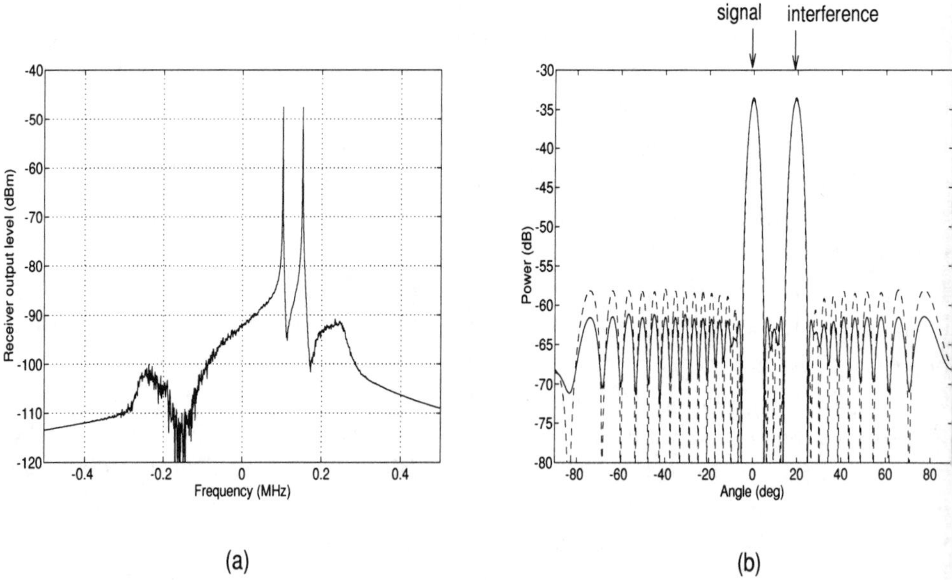

Figure 4.18 Signal-plus-interference case (single-tone interferer, SIR = 0 dB): (a) receiver element output spectrum; (b) array beam pattern. Receiver input level: −70 dBm.

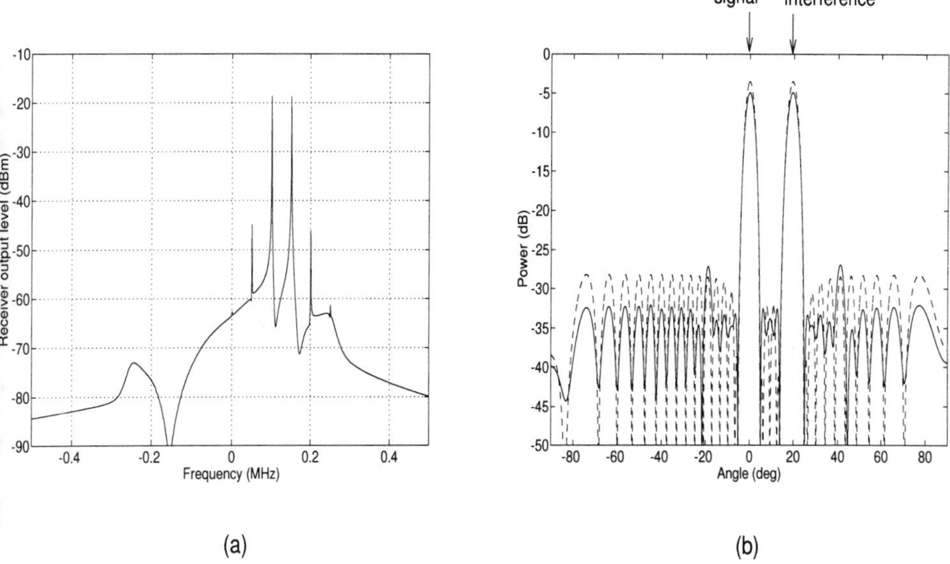

Figure 4.19 Signal-plus-interference case (single-tone interferer, SIR = 0 dB): (a) receiver element output spectrum; (b) array beam pattern. Receiver input level: -40 dBm.

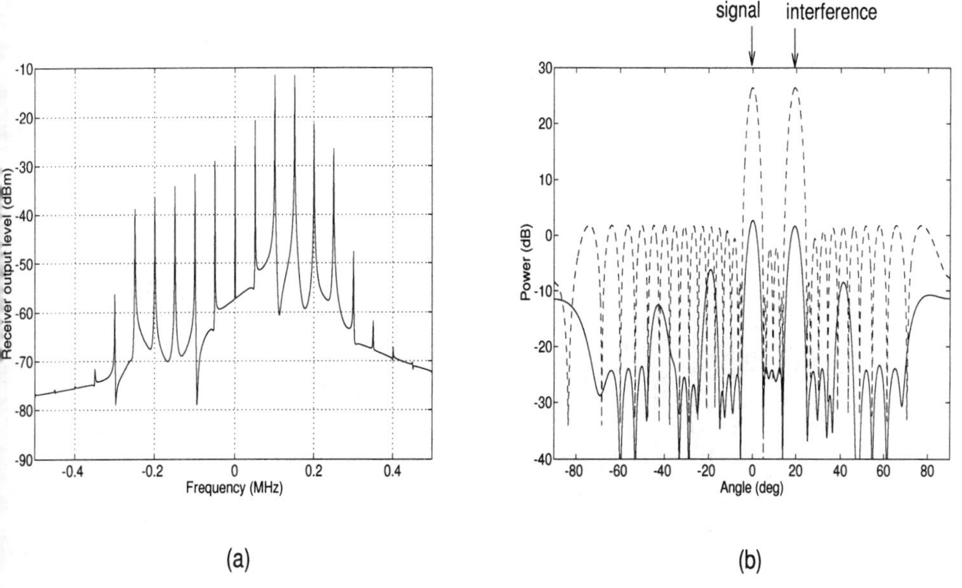

Figure 4.20 Signal-plus-interference case (single-tone interferer, SIR = 0 dB): (a) receiver element output spectrum; (b) array beam pattern. Receiver input level: -10 dBm.

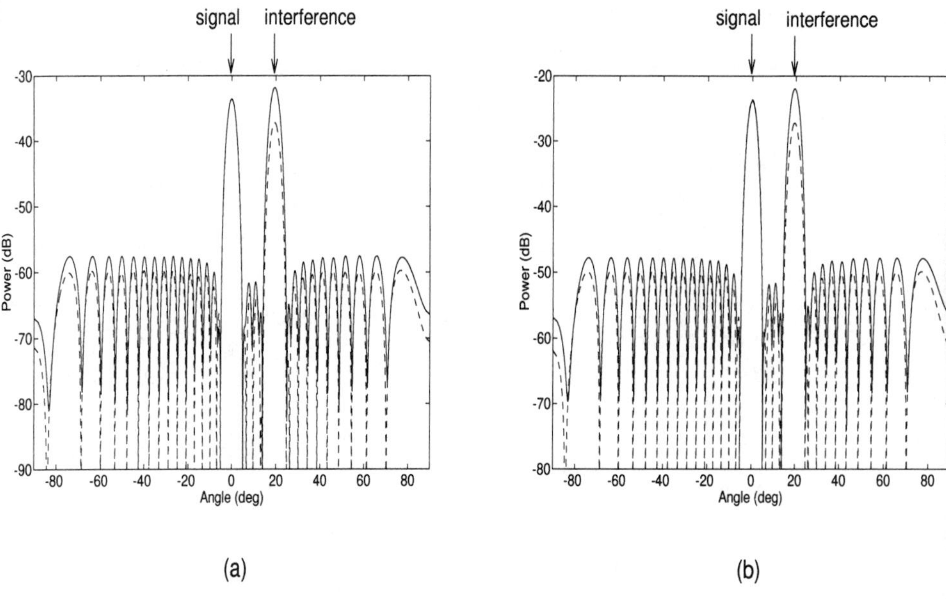

Figure 4.21 Signal-plus-interference case (wideband noise interferer, SIR = 0 dB): (a) −70 dBm; (b) −60 dBm.

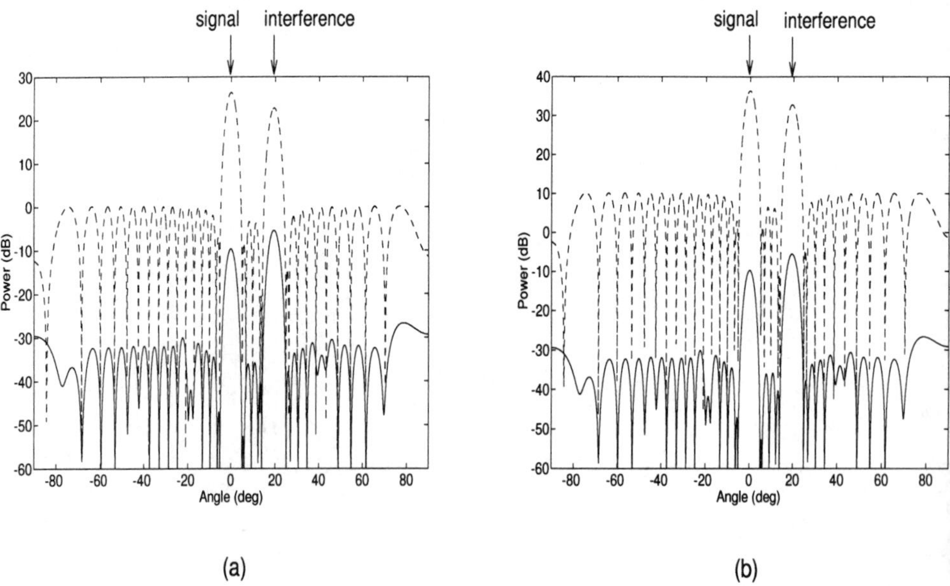

Figure 4.22 Signal-plus-interference case (wideband noise interferer, SIR = 0 dB): (a) −10 dBm; (b) 0 dBm.

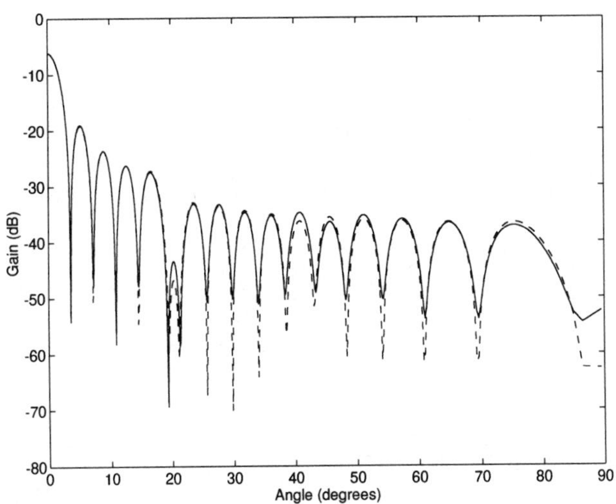

Figure 4.23 Adaptive beamforming patterns for the nonlinear receiver: SNR = 40 dB (dashed curve); SNR = 20 dB (solid curve).

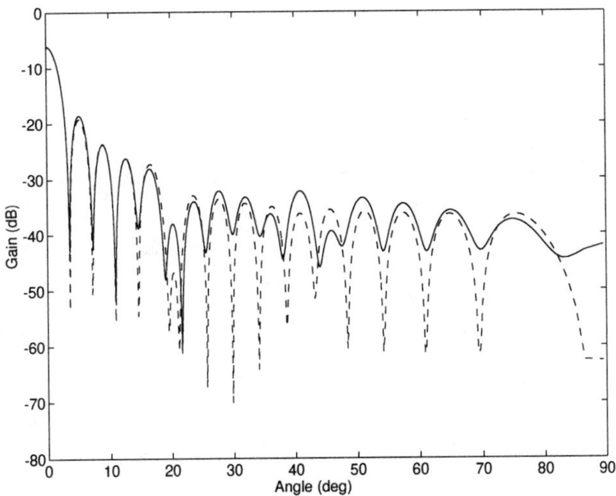

Figure 4.24 Adaptive beamforming patterns for the nonlinear receiver; SNR = 40 dB (dashed curve); SNR = 10 dB (solid curve).

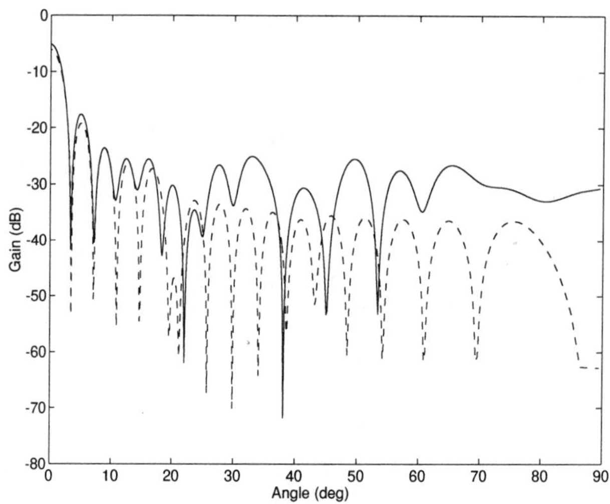

Figure 4.25 Adaptive beamforming pattern for the nonlinear receiver; SNR = 40 dB (dashed curve); SNR = 0 dB (solid curve).

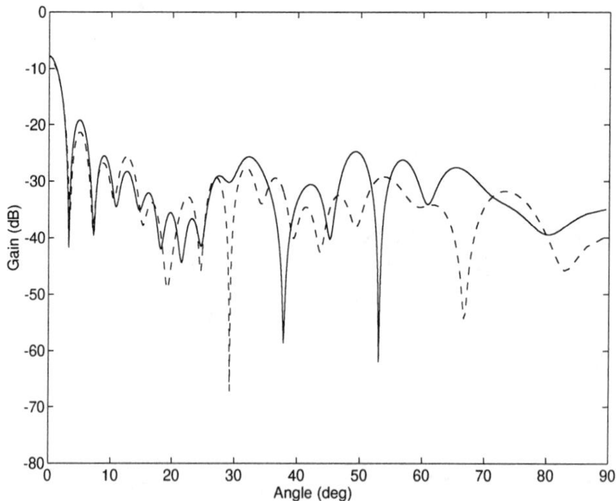

Figure 4.26 Comparison of adaptive beamforming patterns for the linear (dashed curve) and nonlinear (solid curve) receivers: SNR = 0 dB and input level is -20 dBm.

References

[1] L. Eber, "Digital beam steering antenna," Technical Report RADC-TR-88-83, General Electric Company, 1988.

[2] J. G. Proakis and D. G. Manolakis, *Digital Signal Processing*, Macmillan, New York, second edition, 1992.

[3] B. D. Steinberg, *Principle of Aperture and Array System Design*, John Wiley & Sons, New York, 1976.

[4] M. C. Jeruchim et al., *Simulation of Communication Systems*, Plenum Press, New York, 1992.

[5] N. M. Blachman, "Band-pass nonlinearities," *IEEE Trans. Inform. Theory*, vol. IT-10, pp. 162–164, Apr. 1964.

[6] N. M. Blachman, "Detectors, bandpass nonimearities, and their optimization: Inversion of the Chebyshev transform," *IEEE Trans. Inform. Theory*, vol. IT-17, pp. 398–404, July 1971.

[7] A. J. Cann, "Nonlinearity model with variable knee sharpness," *IEEE Trans. Aerosp. Electron. Syst.,*, vol. AES-16, pp. 874–877, Nov. 1980.

[8] R. Blum and M. C. Jeruchim, "Modeling nonlinear amplifiers for communication simulation," in *IEEE Int. Conf. Comm., ICC/89*, Boston, pp. 1468–1472, 1989.

[9] C. Laperle, T. Lo, and J. Litva, "Modelling of nonlinearities and a study of their effects on beamforming," Technical Report CRL Report no. 285, Communications Research Laboratory, McMaster University, Hamilton, Ont., Apr. 1994.

Chapter 5

Spatial Processing in Communications

Wireless telecommunications services are provided in different forms. For example, in satellite mobile communications, communications links are provided by the satellite to the mobile users. In land mobile communications, communications channels are provided by the basestations to the mobile users. In PCS, communications are carried out in microcell or picocell environments, including outdoor and indoor. However, whatever forms they are in, wireless telecommunications services are provided through radio links, where information such as voice and data is transmitted via modulated electromagnetic waves. That is, regardless of their forms, all wireless telecommunications services are subjected to both the influence of the propagation environments and the interference from cochannel signals in a frequency-reuse system.

The most adverse propagation effect from which wireless communications systems suffer is the multipath fading. Multipath fading, which is usually caused by the destructive superposition of multipath signals reflected from various types of objects in the propagation environments, creates errors in digital transmission. One of the common methods used by the communications community to combat multipath fading is the spatial diversity technique, where two or more antennas at the receiver (or transmitter) are spaced far enough apart that their fading envelopes are uncorrelated. There are a number of schemes for combining the diversity signals, which will be discussed in the next section.

In a frequency-reuse system, communication signals that are transmitted at the same carrier frequency in different cells are separated by a spatial distance to reduce the level of cochannel interference. In principle, in order to achieve a high level of frequency reuse and hence high capacity, a large number of cells are required in a region. This is one of the reasons why microcells and picocells have been proposed for PCS. However, the distance between cochannel cells must be large enough that cochannel interference is lower than some acceptable limit. For a given basestation transmission power level, this puts a limit on the number of cells in a geographical area. In order to increase the capacity, additional measures must be taken to combat cochannel interference. For example, optimum combining can be used to reduce cochannel interference. The concept of optimum combining can be considered as the ultimate form of diversity-combining schemes. In optimum combining, signals from different antenna elements are summed to maximize the signal-to-interference ratio at the output of the combiner, which is exactly the process of adaptive beamforming. It is interesting to note that even though the concept of optimum combining was developed in the field of communications and the concept of adaptive beamforming was developed in the field of radar and antennas, they are basically the same. Thus, the terms *optimum combining* and *adaptive beamforming* are sometimes used interchangeably in the literature. In this chapter, we will also discuss the concept of SDMA and how adaptive beamforming can be used to increase capacity.

5.1 DIVERSITY AND COMBINING TECHNIQUES

Spatial diversity is an effective approach to combating multipath fading. It is used at the receiver (or transmitter) of two or more antennas that are spaced far enough apart that their fading envelopes are uncorrelated. If the probability of the signal at one antenna (the outage probability) below a certain level is p, then the probability of the signals from M identical antennas all being below that level is p^M. Thus, combining the signals from several antennas reduces the outage probability of the system. The essential condition for spatial diversity schemes to be effective is that sufficient decorrelation of the fading envelopes be attained.

Switched Diversity

The simplest and most widely used diversity scheme is switched diversity. In this scheme, as shown in Figure 5.1, the system switches between antennas such that only one is used at any one time. The criterion (e.g., a threshold) for switching is often the detection of loss in signal level on the antenna being used. The switching may be performed at RF to avoid the need for a down-converter for each antenna. In switched diversity, the system performance is greatly affected by the threshold level and switching noise.

Selective Diversity

The selective-diversity scheme is more sophisticated than the switched-diversity scheme, in that the selective-diversity system can monitor the signal level on all the antennas simultaneously and select the strongest at any one time, as shown in

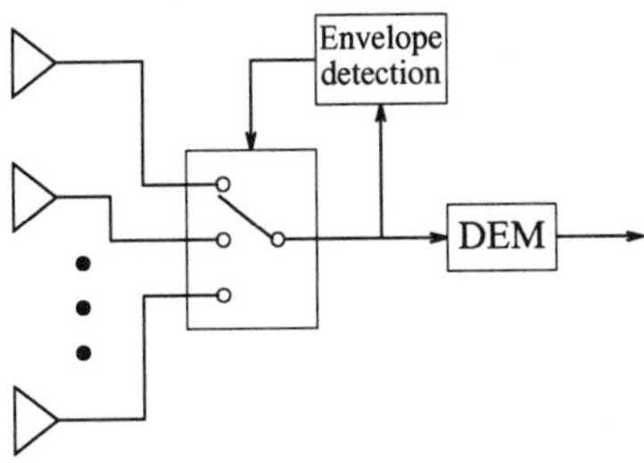

Figure 5.1 Switched-diversity combining.

Figure 5.2. An M-branch selective-diversity receiver needs M receiving front ends, one for receiving the maximum signal and the others for monitoring.

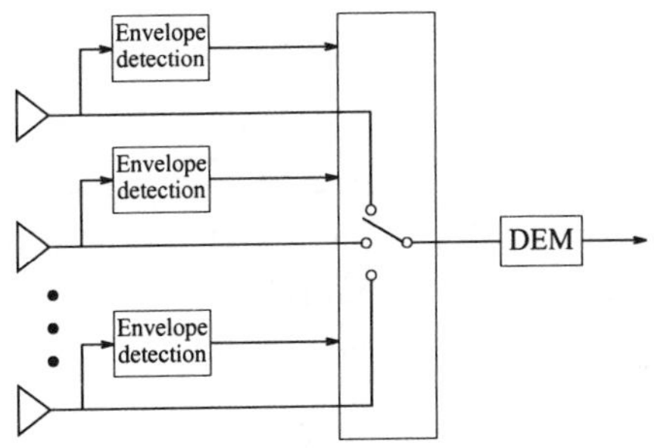

Figure 5.2 Selective-diversity combining.

In the Rayleigh fading environment, the fading of each branch can be assumed to be independent. If each branch has an instantaneous SNR= γ_m, the probability density function of γ_m is given by

$$p(\gamma_m) = \frac{1}{\Gamma} e^{\frac{-\gamma_m}{\Gamma}} \tag{5.1}$$

where Γ denotes the mean SNR at each branch. The probability that a single branch has an SNR less than some threshold γ is given by

$$P\left[\gamma_m \leq \gamma\right] = \int_0^{\infty} p\left(\gamma_m\right) d\gamma_m = 1 - e^{-\frac{\gamma}{\Gamma}} \tag{5.2}$$

The probability that all the branches failing to achieve SNR $= \gamma$ is given by

$$P_M\left(\gamma\right) = P\left[\gamma_1, \gamma_2, \cdots, \gamma_M \leq \gamma\right] = \left(1 - e^{-\frac{\gamma}{\Gamma}}\right)^M \tag{5.3}$$

To evaluate the average SNR of the received signal, we first determine the corresponding probability density function:

$$p_M(\gamma) = \frac{d}{d\gamma} P_M(\gamma) = \frac{M}{\Gamma}\left(1 - e^{-\frac{\gamma}{\Gamma}}\right)^{M-1} e^{-\frac{\gamma}{\Gamma}} \tag{5.4}$$

It follows that the average SNR, $\overline{\gamma}$, is given by

$$\overline{\gamma} = \int_0^{\infty} \gamma p_M(\gamma)\, d\gamma = \Gamma \sum_{m=1}^{M} \frac{1}{m} \tag{5.5}$$

Both switched- and selective-diversity schemes can be easily incorporated into TDMA systems, since the switching can take place in the guard times between user time slots. In systems where transmission is continuous, these two schemes are less appropriate, since switching results in change in channel characteristics such as phase and timing, which in turn may easily result in a burst of demodulation errors whenever switching takes place.

Equal-Gain Combining

The signals on each antenna are combined by bringing all phases of the signals to a common reference point (cophasing). That is, the combined signal is the sum of the instantaneous fading envelopes of the individual branches, as shown in Figure 5.3. The signal envelope at the output of the combiner is given by

$$s_M = \sum_{m=1}^{M} s_m \tag{5.6}$$

If each branch has the same noise power σ_n^2, the total noise power at the output of the combiner is given by

$$\sigma_N^2 = M\sigma_n^2 \tag{5.7}$$

The SNR at the output of the combiner can be expressed as

$$\gamma_M = \frac{s_M^2}{2\sigma_N^2} = \frac{1}{2M\sigma_n^2}\left(\sum_{m=1}^{M} s_m\right)^2 \tag{5.8}$$

The distribution of the squared sum of Rayleigh variables is very difficult to evaluate analytically for $M \geq 3$. Thus, numerical methods are used to obtain the distribution [1]. It has been shown that the distribution curve is slightly below that for the maximum ratio combining technique. The average SNR at the output of the combiner is obtained by simply taking the mean of (5.8); that is,

$$\overline{\gamma}_M = \frac{1}{2M\sigma_n^2}\left\langle\left(\sum_{m=1}^{M} s_m\right)^2\right\rangle = \frac{1}{2M\sigma_n^2}\sum_{l=1}^{M}\sum_{m=1}^{M}\langle s_l s_m\rangle \tag{5.9}$$

since $\langle s_m^2 \rangle = 2\sigma_s^2$, $\langle s_m \rangle = \sqrt{\pi\sigma_s^2/2}$, $\gamma = \frac{\sigma_s^2}{\sigma_n^2}$, and $\langle s_l s_m \rangle = \langle s_l \rangle \langle s_m \rangle$, $l \neq m$ for uncorrelated branch signals, the average SNR is given by

$$\overline{\gamma}_M = \frac{1}{2M\sigma_n^2}\left[2M\sigma_s^2 + M(M-1)\frac{\pi\sigma_s^2}{2}\right] = \gamma\left[1 + (M-1)\frac{\pi}{4}\right] \tag{5.10}$$

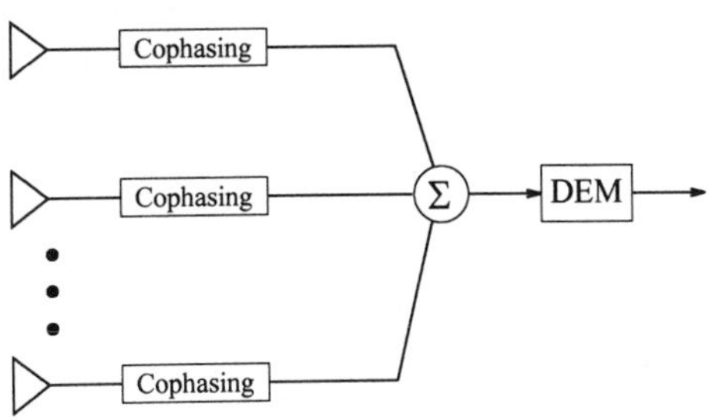

Figure 5.3 Equal-gain diversity combining.

Cophasing is required to reduce the random phase at each branch, which is introduced when the signal propagates from the transmitter to the diversity branch. There are two strategies to carry out cophasing. One is to extract the random phase based on the received signal and the other is to transmit a pilot tone to provide the random-phase information. The extraction of the random phase can be implemented using a feedforward circuit or a feedback circuit, as shown in Figure 5.4.

The random phase can also be reduced using a pilot signal. When transmitting a pilot tone very close to the desired signal carrier, the pilot signal carries the same random phase as the desired signal. Thus, the random phase in the desired signal can be easily canceled out using the pilot signal. However, the frequency separation Δf of the pilot tone and the carrier must be within the coherence bandwidth for a good cancellation [2]; that is,

$$\Delta f < \frac{1}{4\pi D} \tag{5.11}$$

Maximum Ratio Combining

In maximum-ratio combining, instead of using the equal-gain strategy, the signal on each antenna is weighted by its instantaneous carrier-to-noise ratio (CNR), as shown in Figure 5.5. The term *maximum ratio* refers to maximum CNR. The weighting can be carried out in RF or baseband. The weighted signals are combined in the same fashion as in the case of the equal-gain scheme. It is claimed that by weighting the signals, the maximum-ratio combining technique has 1-dB gain in terms of CNR over

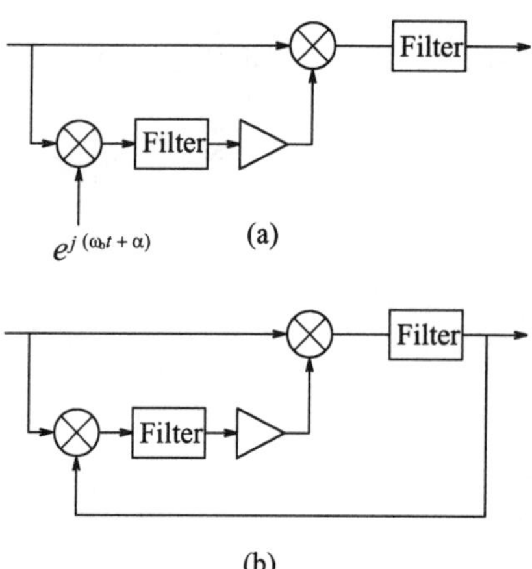

Figure 5.4 Circuits for extraction of the random phase: (a) feedforward cophasing circuit; (b) feedback cophasing circuit.

the equal-gain technique [2]. Furthermore, the maximum-ratio combining technique has been shown to be optimum if diversity branch signals are mutually uncorrelated and follow a Rayleigh distribution [1].

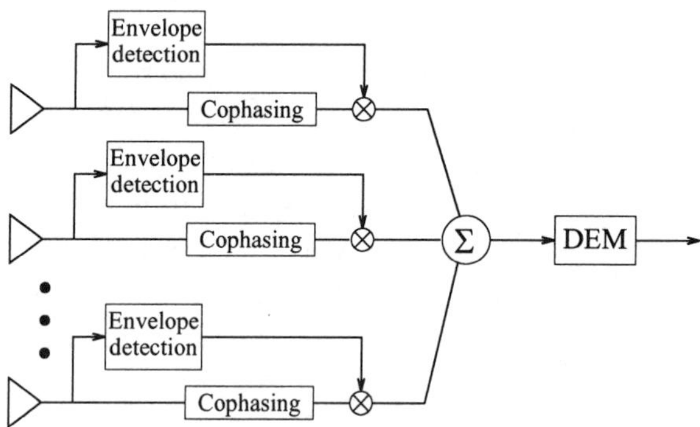

Figure 5.5 Maximum-ratio diversity combining.

If each branch has a gain g_m, the resulting signal envelope at the output of the combiner is given by

$$s_M = \sum_{m=1}^{M} g_m s_m \tag{5.12}$$

Similarly, if each branch has the same noise power σ_n^2, the total noise power at the output of the combiner is

$$\sigma_N^2 = \sigma_n^2 \sum_{m=1}^{M} g_m^2 \tag{5.13}$$

The SNR at the output of the combiner can be expressed as

$$\gamma_M = \frac{s_M^2}{2\sigma_N^2} \tag{5.14}$$

It can be shown that γ_M is maximized when $g_m = \frac{s_m^2}{\sigma_n^2}$, which is the SNR in each branch. It follows that

$$\gamma_M = \frac{1}{2} \frac{\left(\sum_{m=1}^{M} \frac{s_m^2}{\sigma_n^2} s_m\right)^2}{\sigma_n^2 \sum_{m=1}^{M} \left(\frac{s_m^2}{\sigma_n^2}\right)^2} = \frac{1}{2} \sum_{m=1}^{M} \frac{s_m^2}{\sigma_n^2} = \sum_{m=1}^{M} \gamma_m \tag{5.15}$$

Since it has been shown that γ_M follows a chi-square distribution [3], the probability density function of γ_M is given by

$$p(\gamma_M) = \frac{\gamma_M^{M-1} e^{-\frac{\gamma_M}{\Gamma}}}{\Gamma^M (M-1)!} \tag{5.16}$$

The probability that γ_M is less than some threshold is

$$P[\gamma_M \le \gamma] = \int_0^{\gamma} p(\gamma_M)\, d\gamma_M = 1 - e^{-\frac{\gamma}{\Gamma}} \sum_{m=1}^{M} \frac{\left(\frac{\gamma}{\Gamma}\right)^{m-1}}{(m-1)!} \tag{5.17}$$

The average SNR at the output of the combiner is obtained by simply taking the mean of (5.15); that is,

$$\overline{\gamma}_M = \sum_{m=1}^{M} \overline{\Gamma} = M\Gamma \tag{5.18}$$

5.2 SPACE DIVISION MULTIPLE ACCESS

5.2.1 Cell-Based SDMA

As mentioned at the beginning of this book, SDMA is another form of multiple access. Although this term has not seen much use in the open literature, SDMA has in fact been widely used in wireless communications. This may come as a surprise to most readers. For example, in a cellular telephone network, where a large geographical coverage is desired and a large number of mobile transceivers must be supported, the region is divided into a large number of cells. This allows the same carrier frequency to be reused in different cells. In fact, this is a primitive form of SDMA, in that communication signals that are transmitted at the same carrier frequency in different cells are separated by a spatial distance to reduce the level of cochannel interference. In principle, the larger the number of cells in a region, the higher the level of frequency reuse and hence the higher capacity that can be achieved. This is one of the reasons why microcells and picocells have been proposed for PCS. However, the criterion used for defining cochannel cells is that the distance between them is sufficiently large that intercell interference is lower than some acceptable limit. For a given basestation transmission power level, this puts a limit on the number of cells in a geographical area.

In a frequency-reuse system, the term *radio capacity* is usually used to measure the traffic capacity. The radio capacity C_r is defined as [4]

$$C_r = \frac{M}{K \cdot S} \tag{5.19}$$

where M denotes the total number of frequency channels, K denotes the cell reuse factor, and S denotes the number of sectors in a cell. K can be expressed as [5]

$$K = \frac{1}{3}\left(\frac{D}{R}\right)^2 \tag{5.20}$$

where D denotes the distance between two cochannel cells and R denotes the cell radius. In the case of omnicells ($S = 1$ and $K = 7$) the radio capacity is $C_r = M/7$ channels per cell, which is independent of the total number of cells in the system. The corresponding average SIR can be derived based on the propagation path loss of 40 dB per unit distance; that is,

$$C/I = \frac{1}{6}\left(\frac{R}{D}\right)^{-4} = \frac{1}{6}\left(\frac{1}{\sqrt{21}}\right)^{-4} = 18.7 \text{ dB} \tag{5.21}$$

To increase the capacity, the quality, or both, more advanced forms of SDMA are needed. For example, sectorial beams can be used to reduce interference. Figure 5.6 shows two kinds of sectorial cell systems: the seven-cell with three 120° sectors ($K = 7$, $S = 3$) and four-cell with six 60° sectors ($K = 4$, $S = 6$) [6]. In these systems, each sector has a set of unique designated channels. The mobile user moving from one sector or one cell to another sector or cell requires an intracell handoff. The radio capacity and SIR for the 120°-sector system are given as, respectively,

$$C_r = \frac{M}{K \cdot S} = \frac{M}{21} \text{ channels/sector} \tag{5.22}$$

and

$$\text{SIR} = \frac{R^{-4}}{(D + 0.7R)^{-4} + D^{-4}} = 24.5 \text{ dB} \tag{5.23}$$

For the 60°-sector system, the corresponding values are

$$C_r = \frac{M}{K \cdot S} = \frac{M}{24} \text{ channels/sector} \tag{5.24}$$

and

$$\text{SIR} = \left(\frac{R}{D+R}\right)^{-4} = 26 \text{ dB} \tag{5.25}$$

Table 5.1 compares the capacity and SIR for various systems. It should be noted that sectorial cells with intracell handoffs do not increase radio capacity, but increase SIR.

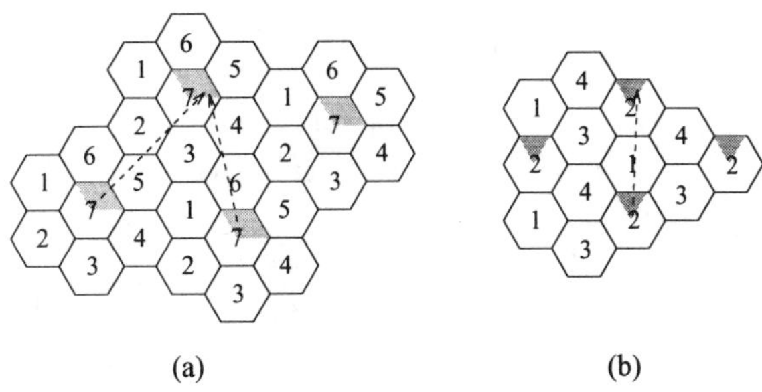

Figure 5.6 Sectorial cells: (a) seven-cell system with 120° sectors; (b) four-cell system with 60° sectors.

5.2.2 Beamforming-Based SDMA

If directional antennas are used, the capacity can be increased. In the case of $K = 7$, each cell has a set of M/K frequency channels. One can use six directional antennas to cover 360° in a cell and divide the whole set of frequency channels that are assigned to the cell into two subsets which are alternating from sector to sector. In this arrangement, there are three cochannel sectors using each subset in a cell, as shown in Figure 5.7. The corresponding radio capacity is given by

$$C_r = \frac{3M}{7 \cdot 6} = \frac{M}{14} \text{ channels/sector} \tag{5.26}$$

which is translated to $\frac{3M}{7}$ channels/cell. That is, the capacity increases by three times as that in the omnicell case. Based on Figure 5.7, there are only two interfering sectorial beams for any sector, and therefore SIR for the worst case is given by

$$\text{SIR} = \frac{R^{-4}}{(3.4R)^{-4} + (5R)^{-4}} = 20 \text{ dB} \tag{5.27}$$

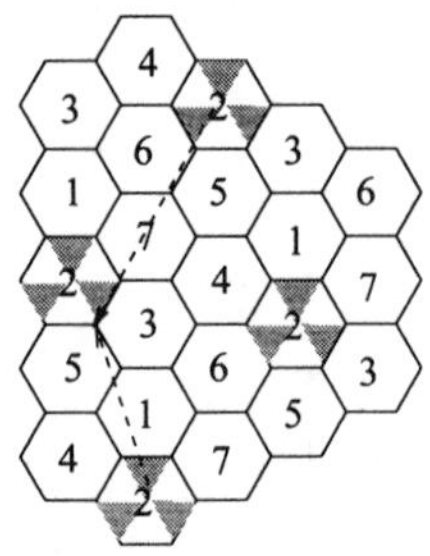

Figure 5.7 60° sectorial beams within a cell in a seven-cell system.

Table 5.1
Radio Capacity and Cignal-to-Interference Ratio in SDMA

	K	S	*Capacity*	C/I
Omnicells	7	1	$\frac{M}{7}$ chs./cell	18 dB
120° sectorial cells	7	3	$\frac{M}{21}$ chs./sector	24.5 dB
60° sectorial cells	4	6	$\frac{M}{24}$ chs./sector	26 dB
60° sectorial beams	7	6	$\frac{3M}{7}$ chs./cell	20 dB (worst case)
N adaptive beams	7	1	$\frac{MN}{7}$ chs./cell	18 dB (worst case)

The ultimate form of SDMA is to use independently steered high-gain beams at the same carrier frequency to provide service to individual users within a cell, as shown in Figure 5.8. That is, a high level of capacity can be achieved via frequency reuse within a cell. To carry out frequency reuse within a cell, a certain level of spatial isolation of cochannel signals is required to maintain an acceptable carrier-to-interference ratio. Adaptive beamforming can provide such a spatial isolation by pointing a beam at the mobile user and at the same time nulling out the interference from cochannel users. Therefore, the spatial efficiency (number of channels per square area unit) can be improved. Since the same frequency or time channel can be used by N mobile users in a cell, the capacity is then given by

$$C_r = \frac{NM}{7} \text{ channels/cell} \tag{5.28}$$

an increase of N times that in the omnicell case. In the worst scenario, where a mobile user in a particular cell is simultaneously interfered with by the beams at the same carrier frequency from the six cochannel cells, SIR is the same as that in the omnicell case; (i.e., SIR = 18 dB). In a normal situation, where the probability for a user to be simultaneously interfered with by the six cofrequency beams is much less than unity and power control is used for the beams, the level of SIR should be relatively higher.

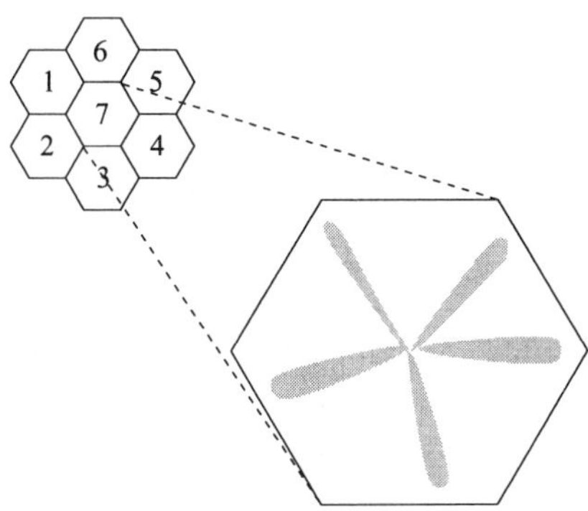

Figure 5.8 Adaptive beamforming for SDMA, showing the use of five independently steered beams at the same carrier frequency to provide service to five users within a cell.

Additional system benefits of applying adaptive beamforming at basestations include:

1. improved immunity to multipath fading because of :
   a. The reduced number of multipath signals, since relatively narrow beams are used;
   b. Optimal diversity combining that is intrinsically carried out by the beamformer.
2. The ability to configure the coverage of each basestation to match the local propagation conditions.

We have seen a number of criteria and algorithms for adaptive beamforming in Chapter 3. They all require some sort of reference signal in their adaptive optimization process. If an explicit reference signal is available in a communications system, it should be used as much as possible for less complexity, high accuracy, and fast convergence. However, an explicit reference signal may not always exist. For this particular reason, a blind adaptive beamforming concept has been conceived and developed. The criteria and adaptive techniques used in blind adaptive beamforming are essentially the same as those used in conventional adaptive beamforming. The key difference is that blind adaptive beamforming does not require an *explicit* reference signal. Instead, a blind adaptive beamforming system generates its own reference signal based on the *implicit* characteristics of the wanted signal. These hidden characteristics include the constant modulus property, information from the decision process, cyclostationarity, and other similar communications signal features. In the next section, we will discuss a few types of blind adaptive techniques.

5.3 BLIND ADAPTIVE BEAMFORMING

5.3.1 Constant Modulus Adaptive Beamforming

Godard was the first to make use of constant modulus (CM) property to carry out blind equalization in two-dimensional digital communications systems [7]. His approach exploits the low modulus variation of most communications signals, such as the signals with frrequency modulation (FM), phase-shift keying (PSK), frequency-shift keying (FSK), or quadrature amplitude modulation (QAM) PCM modulation formats. Although the *Godard algorithm* was developed for equalization, it can be directly apply to blind adaptive beamforming.

As we recall, the output of a DBF array is given by

$$y(n) = \boldsymbol{w}^H(n)\boldsymbol{x}(n) \tag{5.29}$$

Under the assumption that the transmitted signal $s(n)$ has a constant envelope, the array output $y(n)$ should have a constant envelope as well. However, if multipath is present, there is the multipath fading effect and the array output $y(n)$ will not have a constant envelope. The objective of CM beamforming is to restore the array output $y(n)$ to a constant envelope signal, on average. This can be accomplished by adjusting the array weight vector $\boldsymbol{w}$ in such a way as to minimize the cost function, which measures the signal modulus variation and is given in a general sense by

$$\varepsilon = \mathcal{D}\{f[s(n)], f[y(n)]\} \tag{5.30}$$

where $\mathcal{D}$ and f are some defined specific distance metrics. In particular, the Godard algorithm minimizes the nonlinear cost function that takes the form of

$$\varepsilon = E\left[(r_p(n)1 - |y(n)|^p)^2\right] \tag{5.31}$$

where

$$r_p(n) = \frac{E[|s(n)|^{2p}]}{E[|s(n)|^p]} \tag{5.32}$$

and p is a positive integer. In the case where $p = 2$, the algorithm is referred to in the literature as the constant modulus algorithm (CMA), which was proposed ndependently of Godard's paper, by Treichler and Agee [8] to compensate for ading of constant envelope signals. The cost function ε is a positive definite measure f the average amount that the array output $y(n)$ deviates from the unity modulus ondition. The objective is to find a set of values for the array weight vector that vill minimize ε. Of course, there are many possible approaches to finding the ppropriate weight vector. For real-time implementation, a simple search algorithm desirable. For example, the steepest-descent method can be used to minimize ε. n fact, the development of the updating equation is very similar to that in the case f the LMS algorithm. In particular, the updated value of the weight vector at time $+1$ is computed by using the simple recursive relation

$$\boldsymbol{w}(n+1) = \boldsymbol{w}(n) + \frac{1}{2}\mu[-\boldsymbol{\nabla}(E\{\varepsilon(n)\})] \tag{5.33}$$

here the gradient $\boldsymbol{\nabla}(E\{\varepsilon(n)\})$ can be found to be

$$\boldsymbol{\nabla}(E\{\varepsilon(n)\} = -E\left\{\left[r_p(n) - |y(n)|^2\right]\ y(n)\boldsymbol{x}(n)\right\} \tag{5.34}$$

n reality, an exact measurement of the gradient vector is not possible, since this ould require a prior knowledge of $E[|s(n)|^p]$. The strategy is to scale $|s(n)|$ to nity and to use the instantaneous estimate,

$$\boldsymbol{\nabla}(E\{\varepsilon(n)\}) \simeq -\left[1 - |y(n)|^2\right]\ y(n)\boldsymbol{x}(n) \tag{5.35}$$

he weight vector can then be updated as

$$\boldsymbol{w}(n+1) = \boldsymbol{w}(n) + \mu\left[1 - |y(n)|^2\right]\ y(n)\boldsymbol{x}(n) \tag{5.36}$$

y defining an error term as

$$\epsilon(n) = \left[1 - |y(n)|^2\right]\ y^*(n) \tag{5.37}$$

e can express the updating equation as

$$\boldsymbol{w}(n+1) = \boldsymbol{w}(n) + \mu\boldsymbol{x}(n)\epsilon^*(n) \tag{5.38}$$

hich is identical to the updating equation (3.39) in the LMS case. That is, the ractical implication is that hardware configured for the implementation of the LMS lgorithm can be used directly for that of the CMA, except that in this case, the eference signal is generated in the way shown in Figure 5.9. A number of particular roperties of the CMA should be noted [8]:

1. Two conditions may lead to a zero-gradient situation where the algorithm stops adapting. The first is the condition $|y(n)| = 1$, which represents the desired convergence optimum. The second condition is $y(n) = 0$, which also forces the gradient to be zero. However, the second condition does not pose a practical problem for the algorithm because of the following two reasons:

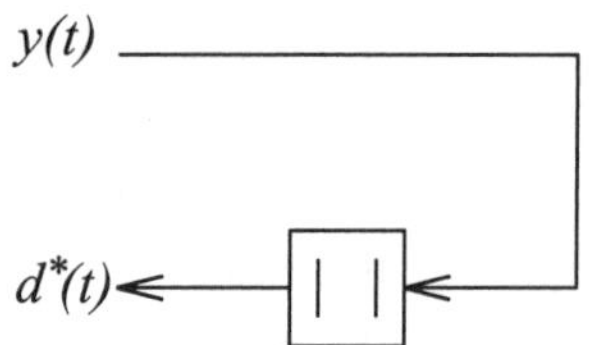

Figure 5.9 Generation of the reference signal in the CMA.

a. The point $y(n) = 0$ is not a stable equilibrium and system noise moves the weight vector $\boldsymbol{w}(n)$ from the zero-gradient point.

b. Given a nonzero snapshot of the array antenna (i.e., $\boldsymbol{x}(n) \neq \mathbf{o}$), the only way to sustain a zero output condition over a large number of snapshots is for the weight vector $\boldsymbol{w}(n)$ to be zero. This condition can be easily detected and corrected.

2. The spatial response characteristics of the beamformer are uniquely defined by the satisfaction of the constant modulus optimization criterion.
3. If the CMA converges, it will converge to the Wiener solution in that the minimization of the constant modulus performance function ε is equivalent to the minimization of the MSE in the output.
4. In the CMA, the wanted signal has been assumed to be of unit amplitude. In general, this would not be the case. In order to minimize the performance criterion, the weights need to be scaled by the adaptive algorithm to adjust the gain to an appropriate level. With the additional control of the amplitude it is more difficult for the algorithm to converge.
5. The convergence of the CMA or Godard's algorithm is not guaranteed because the cost function ε is nonconvex and may have false minima.

5.3.2 Decision-Directed Algorithm

The decision-directed algorithm has been widely used for adaptive equalization to combat intersymbol interference (ISI) in digital communications [9]. In the decision-directed algorithm, the tab weights of the adaptive equalizer are adjusted, via an adaptive process, based on the digital bit stream that is fed back from the hard decision process. Although the decision-directed algorithm has been widely used in adaptive equalization, it is only in recent years that this concept has been applied to diversity combining and beamforming [10–13].

In the decision-directed algorithm, a reference signal is generated based on the outputs of the threshold decision device, as shown in Figure 5.10. The beamformer output, $y(n)$, is demodulated. Based on the demodulated signal, $q(n)$, the decision device makes a decision in favor of a particular value in the known alphabet of the transmitted data sequence that is closest to $q(n)$; that is,

$$r(n) = \text{dec}(q(n)) \tag{5.39}$$

The reference signal, $d^*(n)$, is obtained by modulating $r(n)$. Using the reference signal, one can then establish the cost function for adaptive beamforming; that is,

$$\epsilon(n) = E\left[(d^*(n) - y(n))^2\right] = E\left[\left(d^*(n) - \boldsymbol{w}^H(n)\boldsymbol{x}(n)\right)^2\right] \tag{5.40}$$

The optimum weight vector can be found using the adaptive algorithms that have been discussed in an earlier chapter.

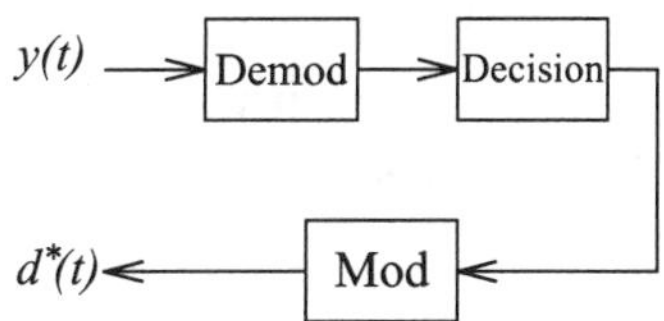

Figure 5.10 Generation of the reference signal in the decision-directed algorithm.

The weight vector in the decision-directed algorithm converges to the Wiener solution in the mean-square sense under the following conditions [14]:

1. The eye pattern is open.
2. The sequence of observations, which is the array output vector $\boldsymbol{x}(n)$ in the case of beamforming, is ergodic in the sense that

$$\lim_{N\to\infty} \frac{1}{N} \sum_{n=1}^{N} \boldsymbol{x}(n)\boldsymbol{x}^H(n) \;\rightarrow\; E\left[\boldsymbol{x}(n)\boldsymbol{x}^H(n)\right] \text{ almost surely} \tag{5.41}$$

The first condition implies that the receiver must be able to lock onto the desired signal. However, the receiver may not always be able to do that. That is, the convergence of the decision-directed algorithm is not guaranteed. This is the same drawback as in the CMA case. The common shortcoming is not surprising because the CMA and decision-directed algorithm belong to a class of blind deconvolution techniques called the Bussgang algorithm [15], where the property of the Bussgang process,

$$E[y(n)y(n+k)] = E[y(n)g(y(n+k))] \tag{5.42}$$

is exploited for blind deconvolution and $g(\cdot)$ is some nonlinear function. In the case of the CMA, $g(y(n)) = |y(n)|$, and in the case of the decision-directed algorithm,

$g(y(n)) = \text{dec}(y(n))$. Due to the nonlinearity of $g(\cdot)$, the cost function of the Buss gang algorithm is nonconvex, and hence the convergence of the Bussgang algorithm is not guaranteed.

5.3.3 Cyclostationary Algorithms

Rather than make use of the constant modulus property of some communication signals, a number of algorithms make use of the cyclostationary properties of certain types of communications signals. The first algorithm that exploits the cyclostation ary properties is called the spectral self-coherence restoral (SCORE) algorithm which was developed by Agee et al. [16], although they termed the cyclostationary properties spectral self-coherence and spectral conjugate self-coherence (hence the name of the algorithm).

Signals are said to exhibit cyclostationarity if their cyclic autocorrelation or cyclic conjugate correlation are nonzero either at some time delay τ or at some frequency shift α. The cyclic autocorrelation of a signal $s(n)$ is defined as

$$R_{ss}^{\alpha} = \langle s(n)s^*(n+\tau)e^{-j2\pi\alpha n}\rangle_{\infty} \triangleq \lim_{N\to\infty} \sum_{n=1}^{N} s(n)s^*(n+\tau)e^{-j2\pi\alpha n} \tag{5.43}$$

and the cyclic conjugate correlation is defined as

$$R_{ss^*}^{\alpha} = \langle s(n)s(n+\tau)e^{-j2\pi\alpha n}\rangle_{\infty} \tag{5.44}$$

When the signal $s(n)$ exhibits cyclostationarity at α_c (i.e., $R_{ss}^{\alpha_c} \neq 0$ or $R_{ss^*}^{\alpha_c} \neq 0$), α_c is referred to as the cycle frequency. The value of α_c for a particular signal is determined by the characteristics of that signal, such as the baud rate and the carrier frequency offset. Therefore, communications signals usually possess their unique, nontrivial cycle frequencies. The commonly used communications signals that possess cyclostationarity include:

1. The double sideband amplitude modulation (DSB-AM) signals, which have cycle frequency at $\alpha_c = 2\Delta f$, where Δf denotes the frequency offset (normalized with respect to the sampling frequency) from the carrier frequency;
2. The binary phase-shift keying (BPSK) signals, which have cycle frequencies at $\{\alpha_c = 2\Delta f \pm kB_r$, for $k = 0, 1, \cdots\}$, where B_r represents the baud rate normalized with respect to the sampling frequency;
3. The binary frequency-shift keying (BFSK) signals, which have cycle frequencies at $\alpha_c = \pm 2\Delta f$;
4. the Gaussian minimum-shift keying (GMSK) signals, which have cycle frequencies at $\{\alpha_c = 2\Delta f \pm k\frac{B_r}{2}$, for $k = 0, 1, \cdots\}$.

The cyclostationarity of signals provides a base for signal selection. That is, a particular signal can be extracted from a signal mixture based on its cycle frequency, provided it has one. For example, we consider a received signal,

$$x(n) = \sum_{l=1}^{L} s_l(n) + \nu(n) \tag{5.45}$$

where $\{s_l(n)\}_{l=1}^{L}$ represent both the desired and the interfering signals, which are mutually independent, and $\nu(n)$ denotes the additive noise. The cyclic autocorrelation of the received signal $x(n)$ can be written as

$$R_{xx}^{\alpha}(0) = \sum_{l=1}^{L} R_{s_l s_l}^{\alpha}(0) + R_{\nu\nu}^{\alpha}(0) \tag{5.46}$$

If only the wanted signal $s_w(n)$ exhibits cyclostationarity at α_w, then we can write

$$R_{xx}^{\alpha_w}(0) = R_{s_w s_w}^{\alpha_w}(0) \tag{5.47}$$

Therefore, the cyclic autocorrelation of s_w can be used to select s_w whether or not it is mixed with other signals in the time domain and/or frequency domain.

There are a number of beamforming techniques that makes use of this selectivity of cyclostationary signals. The most noticeable is the SCORE algorithm for blind adaptive beamforming. The basic concept of the SCORE algorithm is based on the maximization of the cross-correlation coefficient between the outputs of the array $y(n)$ and the reference signal $r(n)$; that is,

$$\rho = \frac{|R_{yr}|^2}{R_{yy} R_{rr}} \tag{5.48}$$

where the reference signal $r(n)$ is not an explicit reference signal as in the LMS case, but is constructed based on the cyclostationarity; that is,

$$r(n) = \boldsymbol{\zeta}^H \boldsymbol{u}^*(n) \tag{5.49}$$

and $\boldsymbol{u}(n) = \boldsymbol{x}^*(n+\tau) e^{j2\pi\alpha_w n}$ is defined as the control signal. That is, to generate the reference signal $r(n)$, the *a priori* knowledge of both the delay and cycle frequency of the wanted signal, τ and α_w, respectively, is required. In (5.49), $\boldsymbol{\zeta}$ can be viewed as the auxiliary weight vector that is designed to extract the cyclic signal. The objective of SCORE is to maximize ρ with respect to $\boldsymbol{\zeta}$ and $\boldsymbol{w}$; that is,

$$\max_{\boldsymbol{\zeta},\, \boldsymbol{w}} \rho = \max_{\boldsymbol{\zeta},\, \boldsymbol{w}} \frac{\left|\boldsymbol{w}^H \boldsymbol{R}_{xu} \boldsymbol{\zeta}\right|^2}{\boldsymbol{w}^H \boldsymbol{R}_{xx} \boldsymbol{w} \boldsymbol{\zeta}^H \boldsymbol{R}_{uu} \boldsymbol{\zeta}} \tag{5.50}$$

In fact, $\boldsymbol{R}_{xu}$ is the cyclic conjugate correlation matrix of $\boldsymbol{x}(n)$; that is,

$$\boldsymbol{R}_{xu} = \langle \boldsymbol{x}(n)\boldsymbol{u}^H(n)\rangle = \boldsymbol{R}_{xx^*}^{\alpha_w}(\tau) \tag{5.51}$$

It is not difficult to see that if ρ is maximized under the constraints that

$$\boldsymbol{w}^H \boldsymbol{R}_{xx} \boldsymbol{w} = 1 \tag{5.52}$$

and

$$\boldsymbol{\zeta}^H \boldsymbol{R}_{uu} \boldsymbol{\zeta} = 1 \tag{5.53}$$

the optimized weight vectors will produce the estimates of the wanted signal $s(n)$ and its time-frequency-shift version $s(n+\tau)e^{-j2\pi\alpha_w n}$; that is,

$$\hat{s}(n) = \boldsymbol{w}^H \boldsymbol{x}(n) \tag{5.54}$$

and

$$\hat{s}(n+\tau)e^{-j2\pi\alpha_w n} = \boldsymbol{\zeta}^H \boldsymbol{u}(n) \tag{5.55}$$

Using the Cauchy-Schwartz inequality, it can be shown that $\boldsymbol{w}$ is optimized, with $\boldsymbol{\zeta}$ fixed, by

$$\boldsymbol{w}_{\text{opt}} \propto \boldsymbol{R}_{xx}^{-1} \boldsymbol{R}_{xu} \boldsymbol{\zeta} \tag{5.56}$$

and similarly, $\boldsymbol{\zeta}$ is optimized, with $\boldsymbol{w}$ fixed, by

$$\zeta_{\text{opt}} \propto \boldsymbol{R}_{uu}^{-1} \boldsymbol{R}_{xu}^H \boldsymbol{w} \tag{5.57}$$

Substituting (5.57) into the objective function yields a generalized Rayleigh quotient in $\boldsymbol{w}$, which can be maximized by setting $\boldsymbol{w}$ to the eigenvector corresponding to the maximum value of the following eigenequation,

$$\lambda_{\text{max}} \boldsymbol{w} = \left[\boldsymbol{R}_{xx}^{-1} \boldsymbol{R}_{xu} \boldsymbol{R}_{uu}^{-1} \boldsymbol{R}_{xu}^H\right] \boldsymbol{w} \tag{5.58}$$

A similar eigenequation can be obtained for $\boldsymbol{\zeta}$; that is,

$$\lambda_{\text{max}} \boldsymbol{\zeta} = \left[\boldsymbol{R}_{uu}^{-1} \boldsymbol{R}_{xu}^H \boldsymbol{R}_{xx}^{-1} \boldsymbol{R}_{xu}\right] \boldsymbol{\zeta} \tag{5.59}$$

It should be noted that the maximum eigenvalue λ_{max} in both eigenequations is the same, representing the maximized objective function value. To extract the wanted signal, $\boldsymbol{w}$ and $\boldsymbol{\zeta}$ can be updated as

$$\boldsymbol{\zeta}(n) = g_\zeta(n) \boldsymbol{R}_{uu}^{-1}(n) \boldsymbol{R}_{xu}^H(n) \boldsymbol{w}(n-1) \tag{5.60}$$

and

$$\boldsymbol{w}(n) = g_w(n) \boldsymbol{R}_{xx}^{-1}(n) \boldsymbol{R}_{xu}(n) \boldsymbol{\zeta}(n) \tag{5.61}$$

where g denotes the corresponding power-normalized gain constant. Up to this point, the correlation matrix $\boldsymbol{R}_{ab}$ of any two signal vectors $\boldsymbol{a}$ and $\boldsymbol{b}$ is understood as an exact measurement (i.e., $N \to \infty$). In reality, an exact measurement is not possible, since N must be finite. Therefore, the estimate of $\boldsymbol{R}_{ab}$ must be used; that is,

$$\hat{\boldsymbol{R}}_{ab} = \frac{1}{N} \sum_{n=1}^{N} \boldsymbol{a}(n)\boldsymbol{b}^H(n) \tag{5.62}$$

which can be updated as

$$\hat{\boldsymbol{R}}_{ab}(n) = (1 - \frac{1}{n})\hat{\boldsymbol{R}}_{ab}(n-1) + \frac{1}{n}\boldsymbol{a}(n)\boldsymbol{b}^H(n) \tag{5.63}$$

The algorithm we discussed here is called the cross-SCORE algorithm, which stems from the fact that the objective function is minimized with respect to both $\boldsymbol{v}$ and $\boldsymbol{\zeta}$. There are variations of the SCORE algorithms. For example, the least-squares SCORE algorithm [16], whose implementation is shown in Figure 5.11, minimizes the least-squares SCORE function,

$$\varepsilon = E\left[|y(n) - r(n)|^2\right] \tag{5.64}$$

the minimization of which leads to the solution for $\boldsymbol{w}$, given as

$$\boldsymbol{w}_{\text{opt}} = \beta \boldsymbol{R}_{xx}\boldsymbol{v} \tag{5.65}$$

where $\boldsymbol{v}$ is the array propagation vector for the desired signal defined by (3.6) and the scalar β is given by

$$\beta = R^{\alpha}_{ss^*}(\tau)e^{-j\pi\alpha\tau}\boldsymbol{v}^T\boldsymbol{\zeta} \tag{5.66}$$

It is interesting to note that (5.65) has the same form as the Wiener solution, and, indeed, it converges to the maximum SIR solution as long as $\boldsymbol{\zeta}$ is not orthogonal to $\boldsymbol{v}^*$ [16].

Another variation of the SCORE algorithm is to maximize the absolute cross-correlation coefficient [17] instead of the normalized coefficient (5.48) in the case of the cross-SCORE algorithm; that is,

$$\max_{\boldsymbol{\zeta},\, \boldsymbol{w}} |R_{yr}|^2 = \max_{\boldsymbol{\zeta},\, \boldsymbol{w}} \left|\boldsymbol{w}^H \boldsymbol{R}_{xu}\boldsymbol{\zeta}\right| \tag{5.67}$$

With the treatment similar to that in the case of the cross-SCORE algorithm, $\boldsymbol{w}$ and $\boldsymbol{\zeta}$ can be updated as

$$\boldsymbol{\zeta}(n) = g_{\zeta}(n)\boldsymbol{R}^H_{xu}(n)\boldsymbol{w}(n-1) \tag{5.68}$$

and

$$\boldsymbol{w}(n) = g_w(n)\boldsymbol{R}_{xu}(n)\boldsymbol{\zeta}(n) \tag{5.69}$$

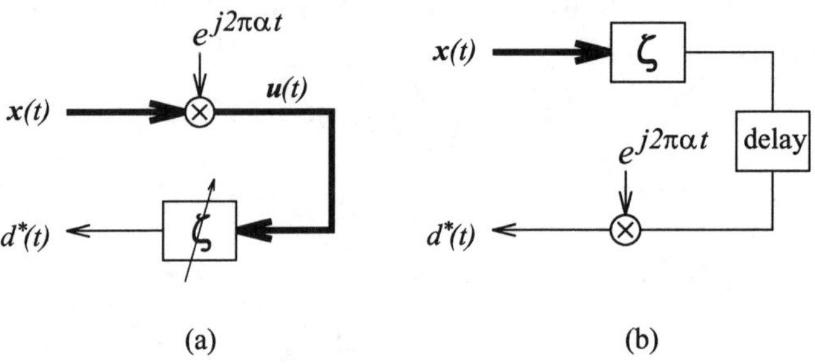

Figure 5.11 Generation of the reference signal in the SCORE agorithms: (a) cross SCORE; (b) least-squares SCORE.

References

[1] W. C. Jakes, Jr., ed., *Mobile Microwave Communication*, John Wiley & Sons, New York, 1974.

[2] W. C. Y. Lee, *Mobile Communications Design Fundamentals*, John Wiley & Sons, New York, 1993.

[3] T. S. Rappaport, *Wireless Communications: Principles and Practice*, Prentice Hall, Englewood Cliffs, NJ, 1996.

[4] W. C. Y. Lee, *Mobile Cellular Telecommunications Systems*, McGraw-Hill, New York, 1989.

[5] V. H. McDonald, "The cellular concept," *Bell Syst. Tech. J.*, vol. 58, Jan. 1979.

[6] W. C. Y. Lee, "Applying the intelligent cell concept to PCS," *IEEE Trans. Veh. Technol.*, vol. 43, pp. 672–679, Aug. 1994.

[7] D. N. Godard, "Self-recovering equalization and carrier tracking in a two-dimensional data communication system," *IEEE Trans. Comm.*, vol. 28, pp. 1867–1875, 1980.

[8] J. R. Treichler and B. Agee, "A new approach to multipath correlation of constant modulus," *IEEE Trans. Acoustic, Speech, and Signal Processing*, vol. ASSP-31, pp. 459–472, Apr. 1983.

[9] S. U. H. Qureshi, "Adaptive equalization," *Proc. IEEE*, vol. 73, pp. 1349–1387, Sept. 1985.

[10] J. A. Henriksson, "Decision-directed diversity combiners—Principles and simulation results," *IEEE J. Sel. Areas Comm.*, vol. SAC-5, Apr. 1987.

[11] R. Gooch and B. Sublett, "Joint spatial temporal equalization in a decision-directed adaptive antenna system," in *Proc. 23rd Asilomar Conf. Signals, Systems and Computers*, Noordwijk, Netherlands, 1989.

[12] V. Kezys and J. Litva, "A versatile intelligent antenna testbed for wireless applications," in *Wireless'95 Proc.*, Calgary, Alberta, 1995.

[13] A. Swindlehurst, S. Daas, and J. Yang, "Analysis of a decision directed beamformer," *IEEE Trans. Signal Processing*, vol. 43, pp. 2920–2927, Dec. 1995.

[14] O. Macchi and E. Ewaeda, "Convergence analysis of self-adaptive equalizers," *IEEE Trans. Inf. Theory*, vol. IT-30, 1984.

[15] S. Bellini, "Bussgang techniques for blind equalization," in *Globecom*, Houston, pp. 1634–1640, 1993.

[16] B. G. Agee, S. V. Schell, and W. A. Gardner, "Spectral self-coherence restoral: A new approach to blind adaptive signal extraction using antenna arrays," *Porc. IEEE*, vol. 78, pp. 753–767, Apr. 1990.

[17] Q. Wu, K. M. Wong, and R. Ho, "A fast algorithm for adaptive beamforming of cyclic signals," *IEE Proc. Pt. F*, vol. 141, pp. 312–318, Dec. 1994.

Chapter 6

Digital Beamforming Configurations

As has been mentioned, the main purpose of carrying beamforming in a communications system is to combat multipath, to extend coverage by the basestation, and/or to reuse frequency channels within a cell. In this chapter we will consider the configurations of digital beamforming networks, as well as the implementation of digital beamforming in a system with a particular multiple-access scheme. In particular, considerations will be given to both uplink (reverse-link) and downlink (forward-link) cases.

6.1 DIGITAL BEAMFORMING NETWORKS

In Chapter 2, we discussed both the concepts of element-space beamforming and beam-space beamforming, as well as the trade-off between the two. In element-space beamforming, the data signals from the array elements are directly multiplied by a set of weights to form a beam at a desired angle. By setting appropriate values for the weighting vectors, one can implement beam steering, adaptive nulling, and beam shaping. In the beam-space beamforming, the outputs from the array elements are first processed by a multibeam beamformer to form a suite of orthogonal beams. The output of each beam can then be weighted and the results are combined to produce a desired output. In this section, we will consider the configurations of both the element-space beamforming networks and beam-space beamforming networks. The discussion will be given at a functional level rather than a component level. Furthermore, we are not particularly concerned with how the weights are generated and assume they are given. The weight generation will be discussed in the next few chapters.

6.1.1 Element-Space Beamforming Networks

As we mentioned in Chapter 2, in element-space processing, one manipulates the signals directly from or to the antenna elements by applying weights to the signals to achieve the desired response. The weights can be pre-determined for fixed beamforming such as beam shaping. In the case of adaptive beamforming, the weights are to be updated with time based on a certain objective or criterion, which may be the same as or similar to those discussed in Chapter 3.

Figure 6.1 shows the structure of an element-space digital beamforming network that is used for receiving. As an example, we consider the case where there are L mobile users using the same frequency (or time) channel. The L user signals at the same carrier frequency are received simultaneously by the N antenna elements in the array. In order to separate the user signals, L uplink beams are required to be formed simultaneously at this particular frequency. To carry out beamforming each of the N signals from the antenna elements is divided into L branches. The outputs of the N dividers are organised into L groups, each of which contains N outputs from each of the N dividers. Each group is multiplied by a set of complex weights $\{w_n^l(t), n = 1, \cdots, N\}$, which represents the lth beam. Ideally, the L sets of weights should be mutually orthogonal (i.e., $(\boldsymbol{w}^k)^H \boldsymbol{w}^l = 0$) to minimize interference between the L user signals. The lth output signal, which can be expressed as

$$y_l(t) = \sum_{n=1}^{N} w_n^l(t) x_n(t) \tag{6.1}$$

needs to be further processed, (e.g. using demodulation) to extract the message. It should be noted that the weights $\{w_n^l(t)\}$ are expressed as time-dependent functions. There are a number of reasons for this:

1. The assigned beam must follow For a user that is moving during a call. That is, the weights that represent this particular beam must be adjusted with time.
2. Even for a still user, the communications environment may change with time. For example, the cochannel interfering sources are moving and the angles of arrival of the multipath signals are changing. Therefore, the weights that represent the beam will adapt to the time-varying environment.
3. Time multiplexing is possible if the processing speed of the beamforming network is sufficiently high. In time multiplexing processing, a number of FDMA frequency channels share a DBF processor. In TDMA, the weights are also required to be time functions so that they are synchronized with the corresponding time frames.

Figure 6.2 shows the structure of an element-space digital beamforming network that is used for transmitting. The downlink process is almost the reverse of the uplink case. In this case, there are L message signals to be transmitted to the L mobile users at the same carrier frequency. That is, L downlink beams are required to be formed simultaneously at this particular frequency. To carry out

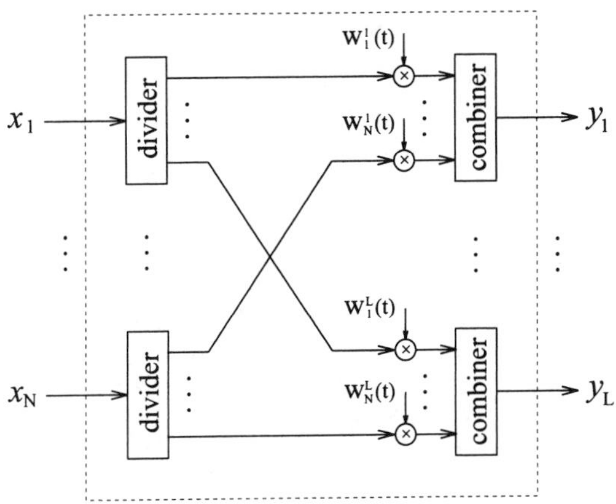

Figure 6.1 The structure of an element-space digital beamforming network used for receiving.

beamforming in the transmitting mode, each of the L signals to be transmitted to the users is divided into N branches, which are multiplied by a set of complex weights $\{w_n^l(t), n = 1, \cdots, N\}$. This particular set of weights represents the lth downlink beam to be formed. Again, it is desirable that the L sets of weights be mutually orthogonal (i.e., $(\boldsymbol{w}^k)^H \boldsymbol{w}^l = 0$) to minimize interference between the L signals that are to be transmitted. The weighted signals are then grouped and combined in the way shown in the figure. The output signal of the nth combiner, which is given by

$$x_n(t) = \sum_{l=1}^{L} w_n^l(t) y_l(t) \tag{6.2}$$

modulates the carrier and the modulated signal is fed to the nth antenna element for transmission.

6.1.2 Beam-Space Beamforming Networks

In beam-space beamforming, we deal with the signals that are represented by beams. As we know from Chapter 2, in order to work with beam-space signals, we need an appropriate transformation between the beam space and the element space. The transformation is usually orthogonal in the sense that the N signals from the antenna elements are represented by N orthogonal signals via the transformation. These orthogonal signals are defined as the orthogonal beams in the beam space. The generation of orthogonal beams can be carried out using a Butler matrix or an

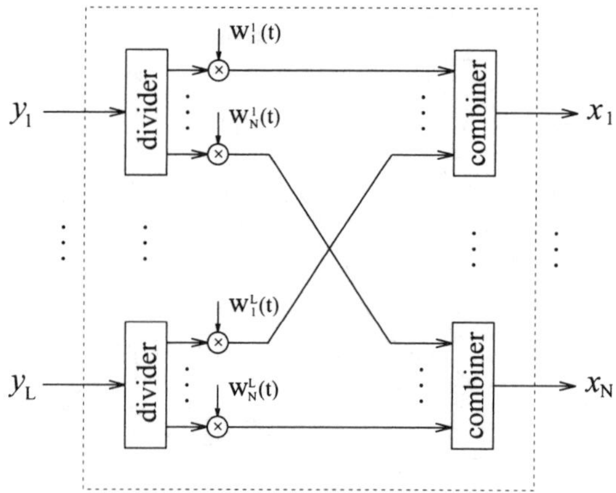

Figure 6.2 The structure of an element-space digital beamforming network used for transmitting.

FFT, as shown in Figure 6.3. A beam-space beamformer is actually a signal processor that weights and combines a set of M ($< N$) orthogonal beams to produce a desired output. The weights can be predetermined for fixed beamforming. In the case of adaptive beamforming, the weights are to be updated with time based on a certain objective or criterion that is the same as or similar to those discussed in Chapter 3. It should be noted that the degrees of freedom used in the adaptive process can be much fewer than those available (i.e., $M \ll N$). This, as we remember from Section 3.4, is referred to as partially adaptive beamforming.

Figure 6.4 shows the structure of a beam-space digital beamforming network that is used for receiving. We again consider the case in which there are L mobile users using the same frequency (or time) channel. That is, the L user signals at the same carrier frequency are received by the N antenna elements of the array. After the orthogonal transformation is applied to the array signals, each user signal may be represented by the weighted combination of one or more orthogonal beams $\{x_n,\ n = 1, \cdots, N\}$. There are a number of reasons why more than one beam is required:

1. Interpolation between the orthogonal beams in order to fine-steer a resultant beam;
2. Phase alignment between beams resulting from multipath signals;
3. Linear combination of a number of orthogonal beams to create nulls in the direction of the interfering signals.

To carry out the beam-space beamforming, x_n is divided into L branches, each

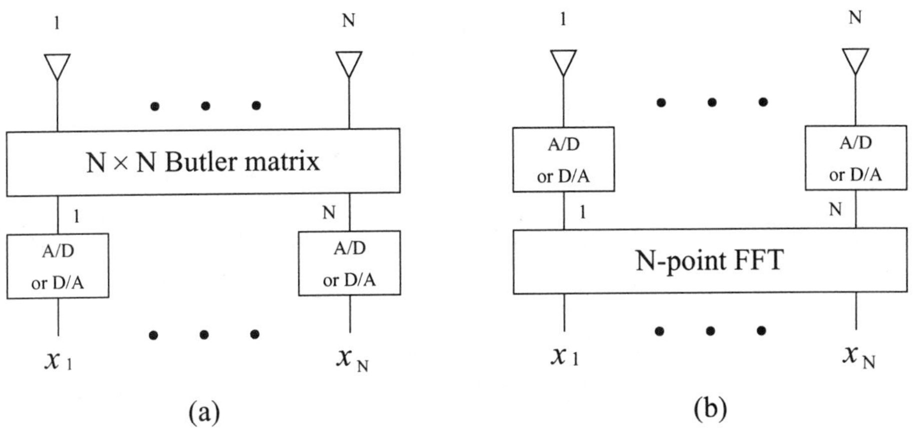

Figure 6.3 Generation of orthogonal beams for beam-spacing beamforming: (a) analog generation using a Butler matrix; (b) digital generation using an FFT.

of which is connected to one of the L beam-select circuits in the way shown in the figure. In order to separate the L user signals, the lth beam-select circuit is instructed by the control signal s_l to choose M appropriate beams among the N orthogonal beams, which are then weighted and combined to represent the lth user signal; that is,

$$y_l(t) = \sum_{m=1}^{M} w_m^l(t) x_{n(m)}(t), \qquad l = 1, \cdots, L \tag{6.3}$$

where the subscript $n(m)$ denotes the index for the M selected beams. For example, for $m = 1, 2$, $n(1) = 5$ and $n(2) = 8$ mean that the fifth and eighth beams are selected. The L beam signals $\{y_l(t), l = 1, \cdots, L\}$ need to be further processed (e.g., using demodulation) to extract the message. It should be noted that the number of orthogonal beams (M) that are required to represent a particular user signal may be different from one user signal to another.

Figure 6.5 shows the structure of a beam-space digital beamforming network that is used for transmitting. The downlink process is almost the reverse of the uplink case. There are L message signals $\{y_l(t), l = 1, \cdots, L\}$ to be transmitted at the same carrier frequency. To carry out beamforming in the transmitting mode, each of the L signals is divided into M branches. The M signals from the lth divider are multiplied by a set of complex weights $\{w_m^l,\ m = 1, \cdots, M\}$ and then assigned to the M desired orthogonal beams by the lth beam-assign circuit that is controlled by the signal s_l. For example, if y_1 is supposedly represented by the weighted fifth and eighth ($M = 2$) beams, y_1 is divided into two branches, each of which will be

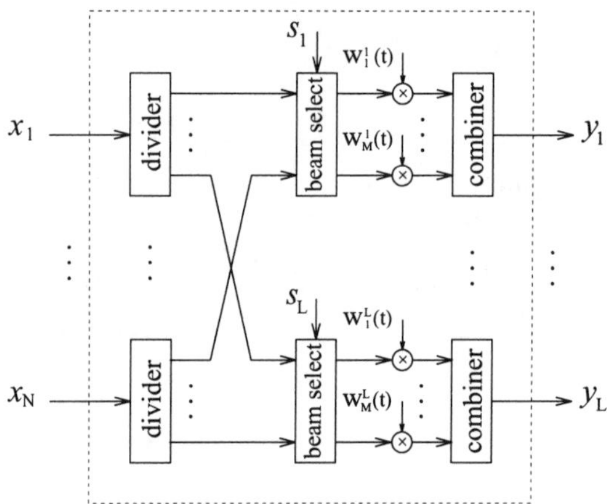

Figure 6.4 The structure of a beam-space digital beamforming network used for receiving.

multiplied by a weight. The weighted signals will then be assigned to the fifth and eighth beams, respectively. The nth beam signal is given by

$$x_n(t) = \sum_{l(n)} y_{l(n)} w_{m(n)}^{l(n)} \tag{6.4}$$

where $l(n)$ denotes the index for the lth message signal that is to be weighted and assigned to the nth beam. The output signal of the nth combiner is fed to the nth input port of the orthogonal beamformer.

6.2 DBF WITH MULTIPLE-ACCESS SCHEMES

In a communications system, a certain form of multiple access must be used in order for users to share the limited channel resource. Therefore, when applied to a wireless communications system, digital beamforming must be applied together with different multiple-access schemes. In this section, we will describe the digital beamforming structures that are suitable for use together with various forms of multiple access. In particular, we will discuss the application of digital beamforming to FDMA, TDMA, and CDMA systems.

6.2.1 DBF with FDMA

In FDMA, the frequency spectrum is divided into slots, each of which is apportioned to a single user within a cell. As we have discussed in an earlier chapter,

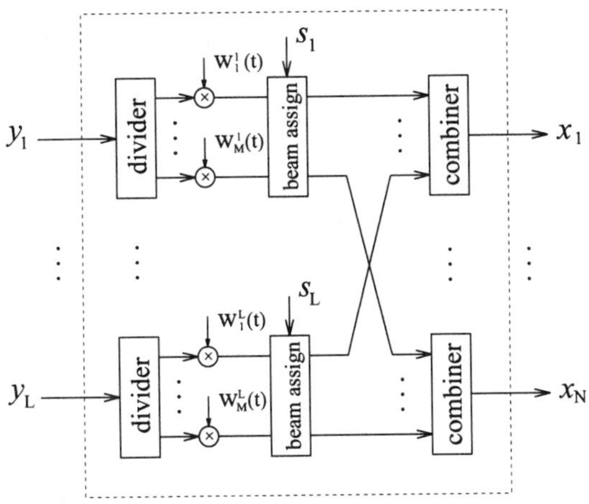

Figure 6.5 The structure of a beam-space digital beamforming network used for transmitting.

when using DBF to implement SDMA, it is possible for multiple users to use the same slot (channel) at the same time. Therefore, DBF must be carried out for individual channels. A number of DBF structures have been proposed for satellite communications systems that use FDMA [1–3]. What is common in these systems is that digital beamforming networks have to be placed after the demultiplexers in the uplink case and before the multiplexers in the downlink case.

The DBF configuration for a basestation system using FDMA in both the uplink and downlink cases is shown in Figures 6.6 and 6.7, respectively, where there are M frequency channels, each of which is used by the system to accommodate L users. That is, the system is able to simultaneously handle $M \times L$ users, at least in theory. The uplink (downlink) system consists of:

1. An N-element antenna array;
2. N receiver (transmitter) modules;
3. N ADCs (digital-to-angalog converters (DAC));
4. N rigital demultiplexers (multiplexers);
5. M deceive (transmit) digital beamforming (RDBF or TDBF) networks;
6. $M \times L$ digital demodulators (modulators).

In the uplink case, beamforming is applied to the signals from all the frequency channels. Therefore, a beamforming network similar to those discussed in the previous section is required for each channel. Beamforming can be implemented in the beam space or in the element space, depending on the configurations of the antenna. If each beamforming network produces L beams, through demodulation, $M \times L$ messages can be recovered. In the downlink case, the process is the reverse of the uplink one. That is, beamforming is applied to the modulated signals from the modulators. The outputs of the beamforming networks are frequency-division multiplexed (FDM) before transmission.

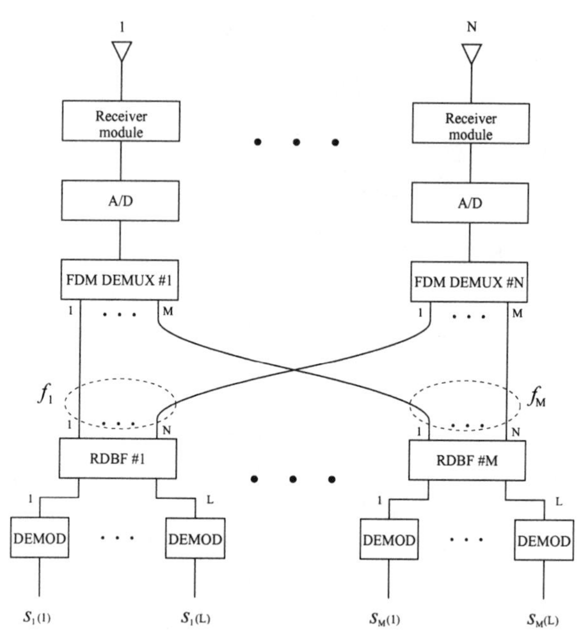

Figure 6.6 Uplink DBF configuration for an FDMA system.

It should also be pointed out that the ADCs are placed immediately after the receivers and the DACs are placed immediately before the transmitters. This arrangement requires only N ADCs for uplink and N DACs for downlink. Furthermore, there is no SNR degradation in processes that follow the analog-to-digital conversion. However, it requires relatively high-speed ADCs and DACs, and high-performance digital processors are needed for frequency demultiplexing and multiplexing. The high-speed and high-performance requirements may be difficult to meet in some applications, such as in a satellite system. The alternative is to carry out the frequency demultiplexing and multiplexing in the analog domain, followed by the analog-to-digital conversion and preceded by the digital-to-analog conversion. This certainly eases the requirement of high-speed ADCs and DACs. However, it requires $N \times M$ ADCs and the analog demultiplexing and multiplexing processes may degrade the SNR of the signals.

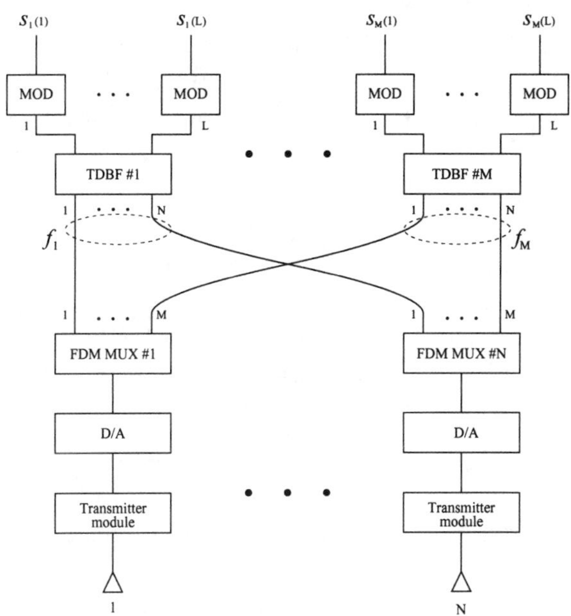

Figure 6.7 Downlink DBF configuration for an FDMA system.

6.2.2 DBF with TDMA

Digital technology makes TDMA a practical access technology. In a TDMA system, each user is apportioned a wideband channel periodically for a brief period of time. The users' transmissions are therefore intermittent in nature, a condition that can only be accommodated by a digital transmitter that can store its source bits and then send them out at a transmission speed higher than that at which they are generated.

The implementation of DBF in a TDMA system is different from that in an FDMA system in that digital beamforming is carried out for the entire frequency band. The DBF configurations for a basestation system using TDMA in both the uplink case and the downlink case are shown in Figures 6.8 and 6.9, respectively. The uplink (downlink) system consists of:

1. An N-element antenna array;
2. N receiver (transmitter) modules;
3. N ADCs (DACs);
4. An RDBF (TDBF) network;
5. L digital demodulators (modulators);

6. L digital demultiplexers (multiplexers).

In the uplink case, beamforming is applied to the received signals from the N antenna elements to produce L uplink beam outputs. Each beam output from the beamforming network will be demodulated and time-demultiplexed into M message signals. That is, the number of simultaneous users can be up to $M \times L$. In the downlink case, the $M \times L$ message signals to be transmitted are arranged into L groups. The M messages in each group are time-multiplexed into a binary stream. A particular digital modulation scheme is then applied to the time-division multiplex (TDM) digital signal. Digital beamforming is applied to the L modulated signals to form L downlink beams. The digital outputs of the beamforming network are converted to analog signals for transmission.

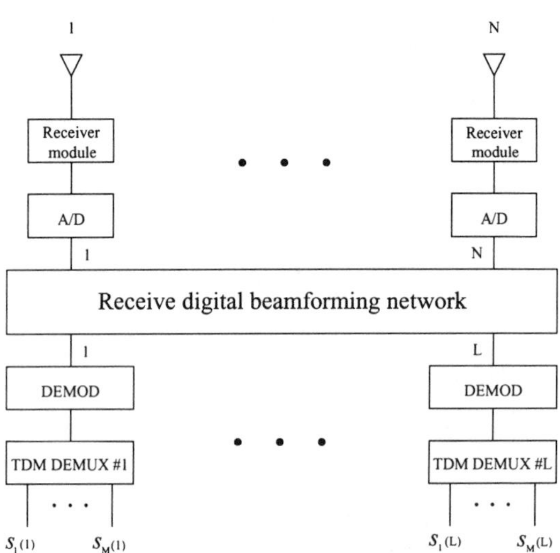

Figure 6.8 Uplink DBF configuration for a TDMA system.

Similar to the FDMA case, the ADCs are placed immediately after the receivers and the DACs are placed immediately before the transmitters. This arrangement requires only N ADCs for uplink and N DACs for downlink. Furthermore, there is no SNR degradation in processes that follow the analog-to-digital conversion. However, it requires relatively high-speed ADCs and DACs. There are alternative configurations that may ease the requirement for high-speed ADCs and DACs. In the alternative configurations, the demodulators, which are placed immediately after the receivers, are followed by the ADCs and the receive beamforming network. Similarly, the modulators, which can be placed immediately before the transmitters, follow the transmit beamforming network and the DACs. In this way, the

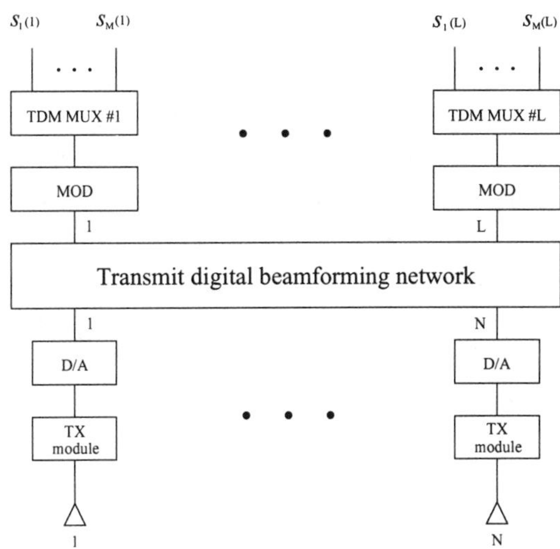

Figure 6.9 Downlink DBF configuration for a TDMA system.

beamforming networks and the ADCs and DACs operate at the bit rate. However, these alternative configurations require that the demodulation and modulation processes be both coherent and linear to preserve the phase information needed for beamforming. They may not be suitable for some systems. For example, GMSK modulation, which is a nonlinear modulation scheme, is used in Global System for Mobile communications (GSM). Thus, the configurations shown in Figures 6.8 and 6.9 must be used if digital beamforming is to be implemented in a GSM system.

The digital beamforming network must process the signals sequentially (i.e., frame by frame). In the downlink case, for example, within a TDMA time frame t_1, the beamforming network produces L beams for the L message signals $\{s_1(1), s_1(2), \cdots, s_1(L)\}$; within another time frame t_2, it produces another L beams for the message signal $\{s_2(1), s_2(2), \cdots, s_2(L)\}$ and so on. There will be problems if the processing time t_p required by the beamforming network is longer than the TDMA time frame t_i. Therefore, it requires a very fast digital beamforming network. An alternative to using a very-high-speed beamforming network is to use a bank of digital beamforming networks, which operate at relatively low speed but in a near-parallel fashion [4]. As shown in Figure 6.10, at the beginning of TDMA frame t_i, the upper switch is at position 1 to load signals $\{s_i(1), s_i(2), \cdots, s_i(L)\}$ into the prebuffer #1. At the beginning of TDMA frame t_{i+1}, the switch at position 2 to load signals $\{s_{i+1}(1), s_{i+1}(2), \cdots, s_{i+1}(L)\}$ into the prebuffer #2 and so on. On the other hand, the lower switch is placed at position 1 after t_p with respect to the beginning of

TDMA frame t_i to send the N processed signals to the DAC. The lower switch is placed at Position 2 after t_p with respect to the beginning of TDMA frame t_{i+1}, and so on. The implementation of this scheme requires additional hardware such as buffers and additional software such as synchronization control. In either the case of a single very fast beamforming network or the case of multiple beamforming networks, the time delay due to beamforming is the processing time t_p.

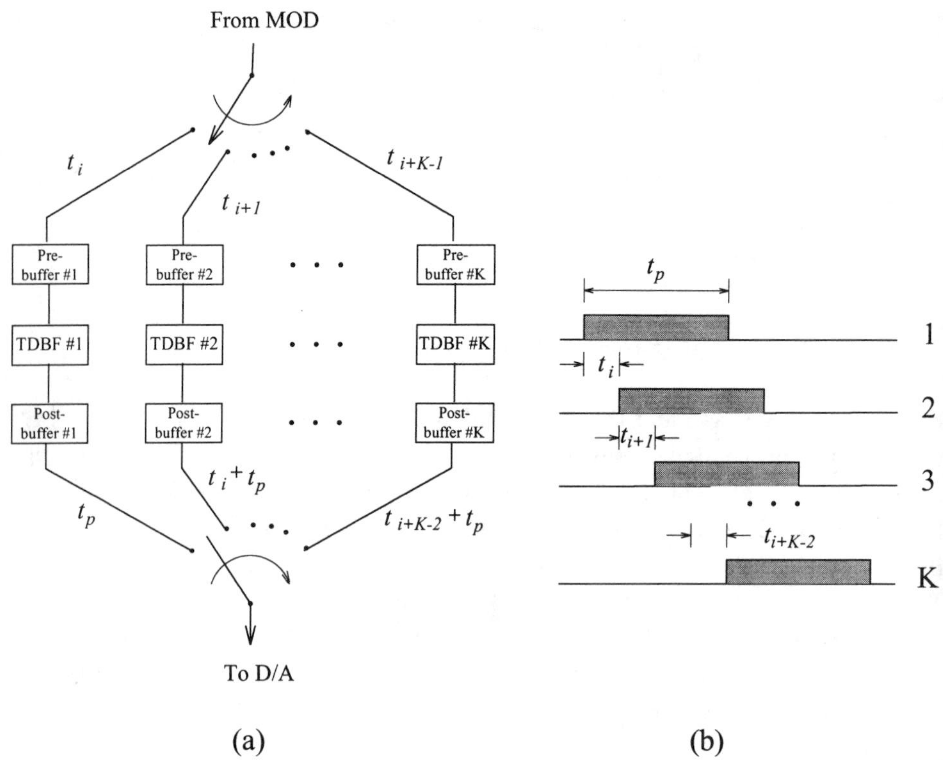

Figure 6.10 Multiple digital beamforming networks for TDMA applications: (a) signal flow structure; (b) time frame structure.

6.2.3 DBF with CDMA

A CDMA system may employ the spread-spectrum modulation; that is, each user's digital waveform is spread over the entire frequency spectrum that is allocated to all users of the network. Each of the transmitted signals is modulated with a unique code that identifies the sender. The intended receiver then uses the appropriate code to detect the signal of his choice.

The implementation of DBF in a CDMA system is different from that in an FDMA system or in a TDMA system. In an FDMA system, beamforming is carried out continuously in each individual channel, whereas in a TDMA system, beamforming is carried out on a frame-by-frame basis for the entire TDMA frequency band. In a CDMA system, beamforming is carried out continuously for the entire CDMA frequency band. Moreover, the choice of DBF configuration depends on the type of CDMA system, namely, a synchronous or asynchronous CDMA system.

In a synchronous CDMA system, the information bit duration of each user signal in the system is time-aligned at the basestation. The DBF configurations for a basestation system in both the uplink and the downlink cases are shown in Figures 6.11 and 6.12, respectively. The uplink (downlink) system consists of:

1. An N-element antenna array;
2. N receiver (transmitter) modules;
3. N ADCs (DACs);
4. M digital demodulator (modulator) banks, each of which consists of N correlators and samplers;
5. N RDBF (TDBF) networks.

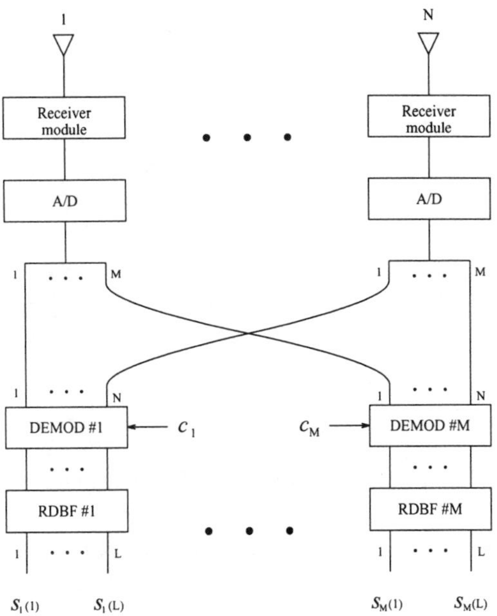

Figure 6.11 Uplink DBF configuration for a CDMA system.

In the uplink case, a beamforming network is required for user signals that use an identical code. This requires that beamforming is carried out following the demodulation process (or code filtering). An example of this type of configuration is given in [5]. If each beamforming network produces L beam outputs, the number of simultaneous users can be up to $M \times L$. In the downlink case, the $M \times L$ message signals to be transmitted are arranged into M groups. Beamforming is applied to the L messages in each group. The beamforming outputs are spread with a particular code. The code-division multiplexed (CDM) signals are combined and converted to analog signals for transmission. In these configurations, the beamforming networks operate at the bit rate. However, it should be noted that the demodulation and modulation processes must be both linear and coherent to preserve the phase information that is required for beamforming.

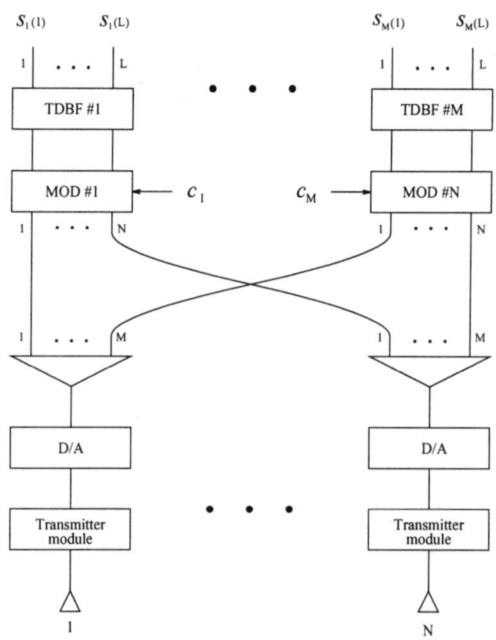

Figure 6.12 Downlink DBF configuration for a CDMA system.

In order for DBF to be implemented in an asynchronous CDMA system, the above configurations must be modified. As shown in Figure 6.13, the receive beamforming networks must be placed before the demodulators and the transmit beamforming networks placed after the modulators. That is, the alternative uplink (downlink) system consists of:

1. An N-element antenna array;
2. N receiver (transmitter) modules;
3. N ADCs (DACs);
4. N RDBF (TDBF) networks;
5. $M \times L$ digital demodulators (modulators), each of which consists of a correlator and a sampler.

In Figure 6.13, the time index t_l for c_m denotes the fact that the information bit duration of the user signals do not have to be synchronized. In these configurations, the demodulation and modulation processes can be linear or nonlinear, and coherent or noncoherent. Furthermore, $c_m(t_l)$ can be used as the reference to carry out adaptive beamforming, an example of which is given in [6]. However, it is required that the beamforming networks operate at least at the chip rate.

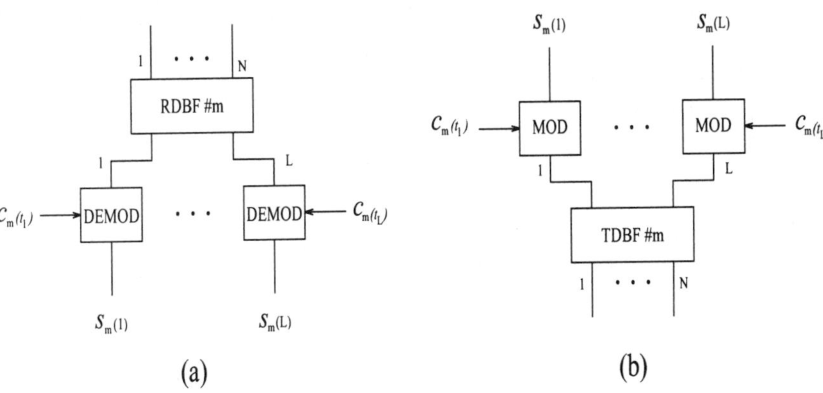

Figure 6.13 Alternative DBF configurations for a CDMA system: (a) uplink; (b) downlink.

References

[1] F. J. Lake and R. P. Curnow, "Active interference suppression and onboard source location within a digital beamforming payload," in *ESA Workshop on Advanced Beamforming Networks for Space Applications*, Noordjwijk, Netherlands, 1991.

[2] M. Barrett, "Digital beamforming network technologies for satellite communications," in *ESA Workshop on Advanced Beamforming Networks for Space Applications*, Noordjwijk, Netherlands, 1991.

[3] A. D. Craig, C. K. Leong, and A. Wishart, "Digital signal processing in communications satellite payload," *Electron. Comm. Eng. J.*, vol. 4, June 1992.

[4] T. Lo, C. Laperle, Y. Shen, M. Zhang, and J. Litva, "Digital beamforming for mobile satellite communications—Technology assessment and design study," Technical Report CRL Report no. 284, Communications Research Laboratory, McMaster University, Hamilton, Ont., Dec. 1993.

[5] A. F. Naguib and A. Paulraj, "Performance of CDMA cellular networks with base-station antenna arrays," in C. G. Gunther, ed., *Mobile communications—Advanced Systems & Components*, Springer-Verlag, pp. 87–100, March 1994.

[6] R. T. Compton, Jr., *Adaptive Antennas: Concepts and Applications*, Prentice Hall, Englewood Cliffs, NJ, 1988.

Chapter 7

DBF in Mobile Satellite Communications

Land mobile satellite communications are evolving towards providing compatibility with the services offered by terrestrial personal communications systems. Satellites can offer services to users in rural and remote areas which are outside the range of terrestrial communications systems. Through the use of mobile satellite (MSAT) communications systems, users anywhere in the world will be able to access voice and data communications while traveling as passengers in airplanes, trains, or onboard ships. In order to cope with the ever increasing demand for mobile satellite communications, the spectral efficiency must be improved. One way to achieve high spectral efficiency is to carry out SDMA using DBF technology. In this chapter, we will discuss the applications of DBF in mobile satellite communications. In particular, the discussions will be centered on the considerations of onboard digital beamforming and various types of DBF antenna systems. A number of system examples are also given in the last section of this chapter.

7.1 ONBOARD DIGITAL BEAMFORMING

In recent years the concept of using DBF technology for applications to future communication satellite payloads [1–9] has been slowly taking root in the technical community. It is envisaged that these satellite systems would support a few thousand to tens of thousands of channels. As we have mentioned earlier, SDMA is a multiple-access scheme that provides communications channels to different areas using the same frequency, which is the frequency-reuse concept. For example, in a mobile satellite communications system where FDMA is used, communications links are provided by antenna beams at the same frequency to different geographical areas. Figure 7.1 shows the situation in which 13 beams in four frequency bands are used to provide service to a region. If TDMA is used, multiple beams are used in each time frame to provide service at the same frequency, as shown in Figure 7.2. In both examples, the system capacity is increased by a factor of $\frac{13}{4}$, compared to the case in which one beam per carrier frequency or per time frame is used.

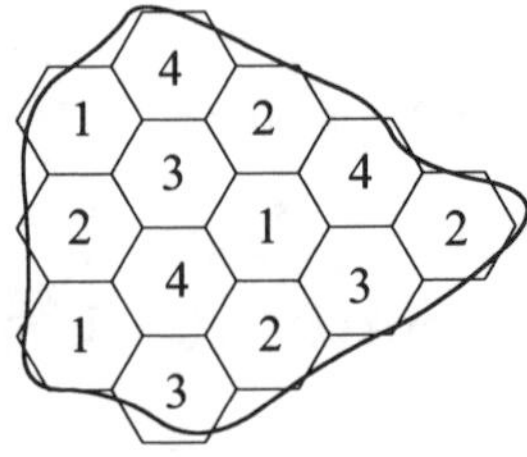

Figure 7.1 Frequency reuse with SDMA, showing the area covered by beams that are arranged to reuse the same frequency bands indicated by numbers.

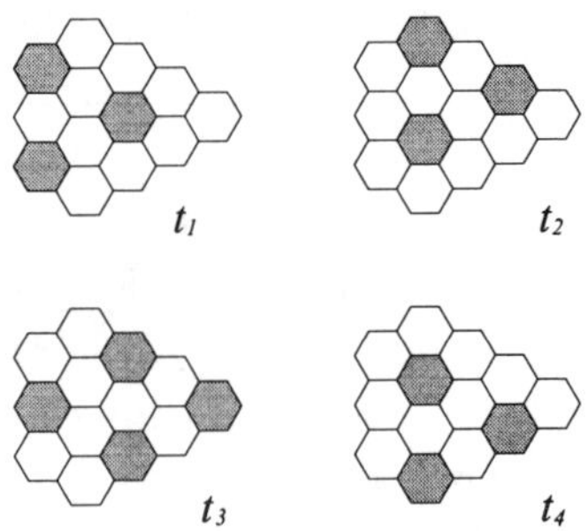

Figure 7.2 Time frame reuse with DBF, showing that in each time frame, more than one beam (shaded hexagons) can be formed to provide communications channels.

7.1.1 Beamforming Strategies

The number of beams that are to be generated to cover a region depends on the physical size of the satellite antenna array and beamforming strategies to be employed. The effective aperture of an array determines its HPBW. There are a number of beamforming strategies. For instance, a single beam may be formed to provide multichannels, which is suitable for multichannel-per-carrier (MCPC) transmission. A single beam may only provide a single channel, which is the case with multichannel-per-carrier (SCPC) transmission. Furthermore, a beam may be formed nonadaptively or adaptively. Let us consider the case of two cofrequency beams. With an SIR requirement of 30 dB, the sidelobe level of a nonadaptive beam must be designed below 30 dB to meet the SIR requirement. Assuming that the sidelobe level is achievable, the minimum beam separation is therefore determined by the main lobe. In the multichannel-per-beam case, a number of mobile users are distributed within the HPBW area of the desired beam, and a number of cochannel users are distributed within the cofrequency beam. The minimum separation of two cofrequency beams is therefore determined by the requirement that the −33-dB point of the wanted beam should coincide with the −3-dB point of the interfering beam, as shown in Figure 7.3(a). This minimum separation is approximately 1.8 HPBW. In the single-channel-per-beam case, we require the peak of the cofrequency beam to be outside the wanted beam region where the power density is no less than 30 dB with respect to the peak. Therefore, the minimum separation is at least 1.2 HPBW, as shown in Fig 7.3(b). If adaptive beams are used, 30-dB nulls can be steered towards the cochannel interferers; and the 30-*dB* sidelobe level requirement may not be necessary. The wanted beam and cofrequency beam may be as close as 0.8 HPBW. Furthermore, it is much easier to achieve a 30-dB null than a 30-dB sidelobe level and a 30-dB sidelobe beam has a larger HPBW than an adaptive beam. As a result, the level of frequency reuse achievable in the adaptive case will be higher, which represents a higher system capacity.

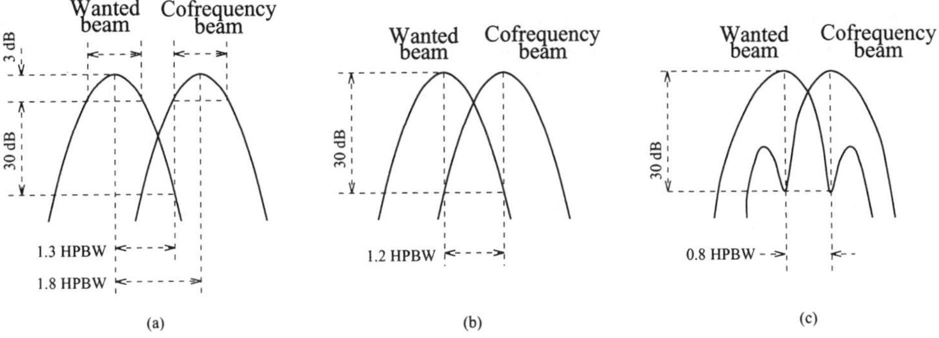

Figure 7.3 Main lobe frequency reuse constraint: (a) nonadaptive multichannel-per-beam; (b) nonadaptive single-channel-per-beam; (c) adaptive single-channel-per-beam.

7.1.2 Acquisition

If adaptive beamforming is to be implemented onboard a communications satellite, a reference signal is required to enable the satellite to steer a beam towards a mobile user. As we pointed out in Section 3.5, the reference can be either the location of the mobile user or some type of reference signal sent by the user. If mobile location information is to be used, there are a number of schemes to determine the mobile locations [7]:

1. A mobile user specifies his location to the network control center (NCC) or the satellite system by encoding the location information in the call request signal. The satellite system will decode the location information and its identification and assign a beam to this particular user. This approach requires the mobile user to have a reliable and accurate means to determine the mobile location. With this option, cost and complexity of the mobile terminal may increase due to the additional requirement for determining its own location. Furthermore, the accuracy in the location of the mobile is an important issue if the gain loss due to mispointing is to be minimized. One way to meet the requirements is for the user to use a GPS (Global Positioning System) device. This approach is a simple and fast solution to the mobile location problem. It requires fewer computations and hence less hardware and weight. This option also makes it easier for the subsequent adaptive processing, in that the direction of the desired signal needed as an optimization constraint is readily available to the adaptive processor. The downside of this option is that costly and complex mobile terminals are required.
2. The satellite system may determine the locations of the mobile users using fixed-beam detection. Obviously, this is the only available option for the systems that use fixed beams, such as the single-reflector focal-fed (SRFF) antenna systems (see next section). The location of a mobile is determined by estimating the direction of arrival of the incoming call request signal by detecting the highest power level among the fixed beams. The mobile user identification is also extracted from the call request signal. This option requires relatively low cost, mass, and complexity for both the satellite and the mobile terminals. These are the most desired characteristics when system design is considered. However, this option offers very limited accuracy, which compromises the SIR performance, since inaccuracy in location can cause mispointing of both the antenna beam towards the user and the nulls towards the interfering users. This further reduces the frequency-reuse factor, making it undesirable if the SIR and the frequency-reuse factor are critical parameters for the system.
3. The satellite system may determine the locations of the mobile users using advanced signal processing techniques. This approach requires additional onboard signal processing power. High-resolution spectral analysis techniques will have to be used to meet the accuracy requirements. However, since these

techniques require matrix computation such as matrix inversion or eigen-decomposition, powerful digital signal processors are required onboard the system. This approach may provide sufficient accuracy at the expense of complexity and cost. It also requires additional hardware and power.

If a reference signal is used, the mobile location information is not required. The reference signal may be sent by the mobile user along with the call request signal to the satellite system. The reference can be a tone signal, a spread-spectrum code (e.g., PN sequence), the preamble of a data packet, or other distinct signal characteristics. Some blind beamforming techniques may be suitable, as well. Using a reference signal, the DBF system will steer a beam towards the corresponding user.

7.1.3 Weight Generation

In addition to the requirement for setting the complex weights to steer the beam towards a mobile terminal, the complex weights can be adjusted to suppress cochannel interference. There are three options for generating the beamforming weights: the table-lookup approach, the partial-optimization approach, and the adaptive-processing approach:

1. The table-lookup approach is a nonadaptive approach, in whcih the beamforming weights are designed to steer the main beam towards the mobile and at the same time generally limit the level of sidelobes without invoking an interference suppression function. This will require information on the location of the mobiles. The control of the sidelobe level is carried out by applying an appropriate amplitude taper across the array aperture, with a consequent increase in beamwidth and a decrease in directivity gain. In addition, relatively low sidelobe levels may not be easy to achieve in practical situations. Nevertheless, this approach is simple and fast because weights can be retrieved by table lookup from the memory bank, which stores the precalculated weights.
2. In the partial-optimization approach, signal processing is carried out to determine the weights for horns within each cluster so that the carrier-to-noise ratio is maximized. This approach is especially suitable for the SRFF antenna system, in which a beam is formed using a cluster of horns. However, the processing is only partially optimal in that the system only makes use of a small fraction of the available degrees of freedom.
3. In the adaptive-processing approach, a fully adaptive algorithm is used to generate optimal weights. With this approach, one can dispense with the need for incorporating a low sidelobe design into the antenna beamformer. This means that antenna gain is not subjected to a loss due to aperture taper.

The table lookup approach requires virtually no real-time computation of the weights, since adaptive processing is not implemented. In the other two approaches, signal processing requires a certain number of degrees of freedom in order to properly optimize the weights. The greater the degrees of freedom, the easier it is to achieve the desired optimal weights. On the other hand, a large number of degrees of freedom requires a very heavy computational load. The partial-optimization approach requires a few degrees of freedom, and therefore the computational load is relatively light, but it offers a very limited level of adaptivity. The adaptive-processing approach makes use of all the available degrees of freedom. In terms of optimality, it is desirable, but it requires a powerful DSP processor.

7.2 DBF ANTENNA SYSTEMS

In this section, we consider four antenna configurations that are suitable for digital beamforming satellite antennas: (1) SRFF multiple-beam system, (2) the single-reflector with hybrid transform (SRHT) system, (3) the dual-reflector array-fed (DRAF) system, and (4) the directly radiating array (DRA) system.

7.2.1 The SRFF Multiple-Beam Systems

The SRFF multiple-beam system using DBF is shown in Figure 7.4. The antenna system, which is widely used for generating shaped beams for satellite communications, consists of a reflector illuminated by an array of feed elements. Each feed element generates a constituent beam. A shaped beam can be formed from a number of the constituent beams by using the principle of superposition. This beam concept is the basis for beam-space beamforming.

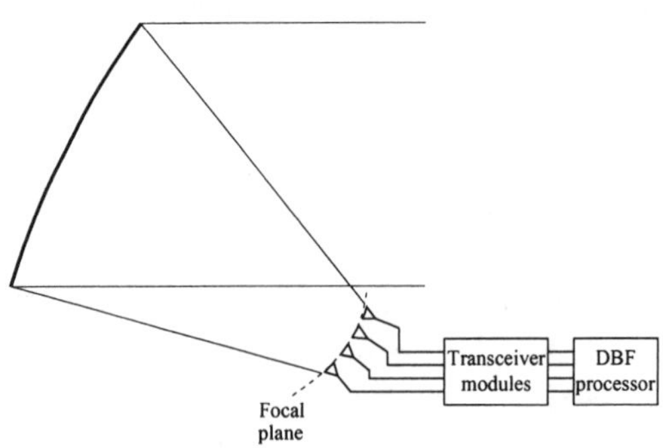

Figure 7.4 An offset SRFF multiple-beam antenna system.

Aperture blockage is avoided by using an offset reflector configuration. However, as a result of the asymmetry of the offset configuration, a cross-polarized component is introduced into the radiation field when the reflector is illuminated by a linear polarized field. When circular polarization is employed, the antenna beam is squinted from the electrical boresight. Thus, difficulties may arise in systems where frequency reuse is based on polarization.

The operation of the reflector on the field distributed over its aperture is analogous to that of a Fourier transformer. That is, the energy is focused or transformed into a beam with a shape that can be described by a sinc function ($\frac{\sin x}{x}$). The reflector produces a spot beam corresponding to a particular feed element located in the reflector's focal region. A beam can be used for both receiving and transmitting energy. Detailed characteristics of an SRFF antenna system can be found in [10].

Due to reflector parallax, only a very limited number of the feed elements located on the focal plane can be used to produce high-performance beams. They are limited to those elements that are close to the reflector's axis of symmetry. Scanning a beam in the far field is accomplished by modifying the amplitude distribution among a cluster of feed elements. For example, if a triplet cluster (containing three horns) is used, coverage beams with different contours can be obtained by appropriately weighting the outputs of the three horns. A simple case is demonstrated in Figure 7.5. The solid curves shown represent the directive gain contour corresponding to the excitation of a single feedhorn. The dashed lines correspond to the simultaneous and equal excitation of two or three adjacent feeds. Each of the contours represents a constant directive gain.

7.2.2 The SRHT System

In the SRFF multiple-beam system, the signals at the array elements are presented in terms of beams. Therefore, it is natural for the digital beamforming process to be carried out in beam space. However, this requires the transceivers to be highly reliable. If any one of them fails, the system performance will be severely degraded. In order to cope with this problem, a Butler matrix or a hybrid transformer can be used to distribute the energy evenly. The configuration is shown in Figure 7.6, where a Butler matrix is placed in between the feed array and the receiver system. Furthermore, adding a Butler matrix can increase the angular scan range of the antenna system [11,12].

It is well known that the relationship between the aperture fields of a parabolic reflector and its focal plane fields is a Fourier transform. For an off-axis wave, the same condition is maintained, provided that the focal point is moved off the axis so that the total angle subtended by the reflector as measured at the focal point is

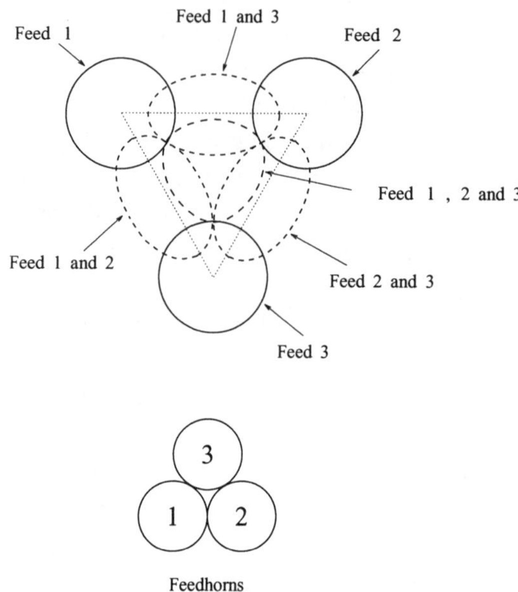

Figure 7.5 Feed and directive gain contours for excitation of the corresponding feeds.

equal to 2θ [11, 13], where θ is the maximum half angle subtended by the reflector from the geometric focus. With this choice for a new focal point, the new focal plane is defined to be normal to the center line of the focal cone of width 2θ, as shown in Figure 7.7. The locus of feed follows the circle passing through the true focus and has a diameter given by [11, 13]

$$\frac{d}{\sin 2\theta} = f\left[1 - \left(\frac{d}{4f}\right)^2\right]\sec^2\theta \tag{7.1}$$

for an aperture of length d in the plane of scan. In this plane, the focal field variation with angle is always a Fourier transform of the aperture plane. Therefore, if another Fourier transformer is placed at the new focal plane, the outputs of the transformer will have a uniform amplitude and a phase variation that is the complex conjugate of the aperture plane field. The second Fourier transform can be performed by a Butler Matrix.

7.2.3 The DRAF System

In the DRAF system, a subreflector, which is confocal and coaxial with the primary reflector, is added to form a dual-reflector system, as shown in Figure 7.8.

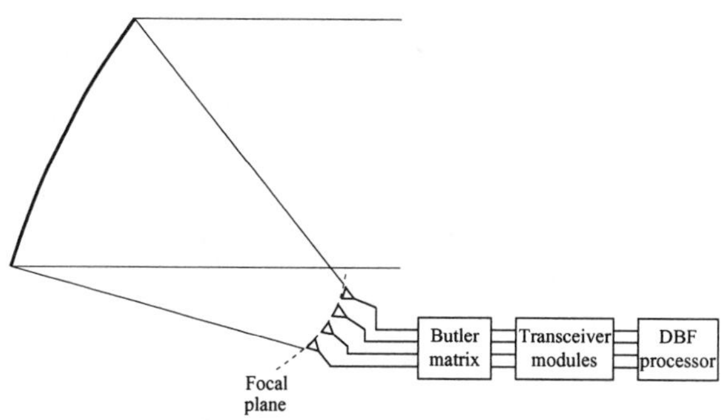

igure 7.6 Single-reflector antenna system with a Butler matrix.

he subreflector, existing in the near field of the array, performs a second Fourier ansform to produce a uniform amplitude field output with a phase variation being ıe complex conjugate of the aperture plane field. Examples of DRAF systems used space applications can be found in [14–16], although phased arrays were used in ıese cases.

Array magnification may be obtained with the confocal arrangement of the two fset parabolic reflectors. A scanned beam is formed by varying the phase and nplitude of elements in the array. The system magnification factor is

$$\mathcal{M} = \frac{D_0}{D_1} = \frac{f_0}{f_1} \tag{7.2}$$

here f_0 and f_1 are the axial focal lengths of the primary reflector and subreflector, spectively. Thus, by choosing

$$\frac{f_0}{f_1} \gg 1 \tag{7.3}$$

small array diameter D_1 is sufficient to intercept all the incident rays. On the ray plane, Σ_1, the center of illumination is determined by the center ray reflected center C_0 of the paraboloid. The array center C_1 must therefore intercept this y. Because C_0 and C_1 are conjugate points, the center of the illumination remains xed even though the direction of the incident wave may change so that the ray cident at C_0 makes a small angle θ_0 with respect to the central ray. For small lue of θ_0, all rays reflected at C_0 pass through C_1 after the second reflection. If $_0$ and Σ_1 are the two planes orthogonal to the center ray through C_0 and C_1, Σ_0 d Σ_1 are optically conjugate planes in the vicinity of C_0 and C_1, and therefore e field E_1 on Σ_1 is the image of the field E_0 on Σ_0. This type of arrangement has number of important characteristics:

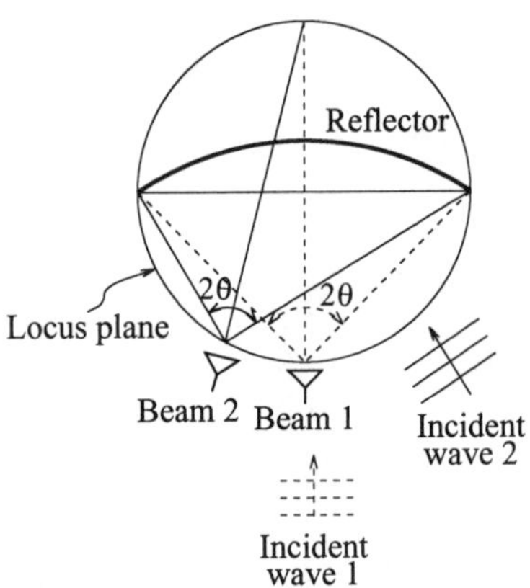

Figure 7.7 Locus of transform plane in which the focal field variation with angle is alway the Fourier transform of the aperture plane.

1. Any small deformation in the main reflector can be easily corrected becaus the main reflector and the array are optically conjugate elements.
2. The transformation that relates the field over the array aperture to the fiel over the main reflector aperture is essentially frequency independent, an therefore it can be approximated by its asymptotic behavior at high frequenc That is, the transformation can be determined accurately using the laws c geometric optics.
3. Grating lobes can be reduced by ensuring that the grating lobes from th array miss the reflector.

The relationship between the angle θ_0 and the angle θ_1 is given by

$$\theta_1 = \mathcal{M}\theta_0 \qquad (7.4$$

For example, if a beam is to be steered at $\theta_0 = 0°$ and $\mathcal{M} = 2.5$, then the angl at the array is θ_1 is equal to 2.5°. It is desirable to have a large $\mathcal{M}$ in order t make the array physically small and at the same time provide sufficiently narro spot beams. However, a large value of $\mathcal{M}$ may result in a higher scan loss, a highe sidelobe level, and a larger aberration, since the array has to scan a larger angl $\mathcal{M}\theta_0$ [16].

For a small value of $\mathcal{M}$, beam degradation can be minimized, but the subreflec tor size may become impractically large for a given primary reflector size. Therefor $\mathcal{M}$ must be chosen carefully so as to achieve a balanced design. Scan losses ca also be reduced by increasing the number of elements for a given aperture size. Un fortunately, this will increase the complexity, mass, and power associated with th

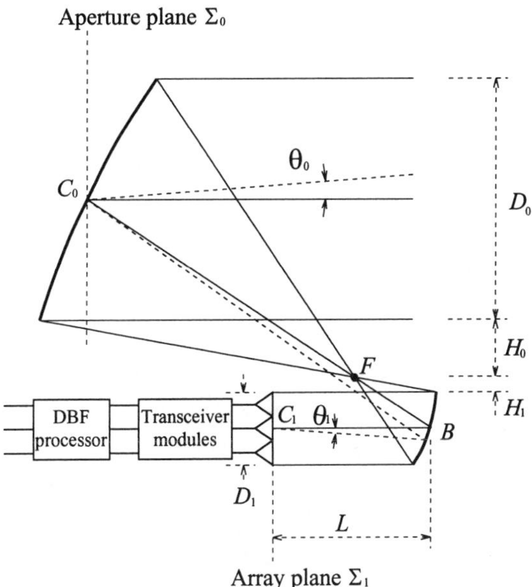

Figure 7.8 Dual-reflector array-fed antenna system.

antenna array. Mutual coupling effects between elements within the array become greater as element spacing decreases.

The approximate formulas relating HPBW and peak directivity D_p to array diameter d are given as

$$\text{HPBW} = \frac{1.02\lambda}{d} \quad \text{(rad.)} \tag{7.5}$$

and

$$D_p = 10\log(60.7f^2d^2) \quad \text{dB} \tag{7.6}$$

where f is frequency in gigahertz (GHz), λ is wavelength in meters, and d is in meters.

The distance between the array and the center of the subreflector L can be adjusted to achieve the maximum illumination efficiency for the array. It can be found that the optimal location of the array is a function of the separation between the center of the primary reflector and the array. A good estimate of the array's location, under limited scan conditions, is given by

$$L = |FB|\frac{\mathcal{M}+1}{\mathcal{M}} \tag{7.7}$$

where $|FB|$ is the radial distance from the focal point to the center of the subreflector.

Since a large offset in a parabolic reflector system produces a strong cross-polarized component in the field, the lowest offset should be chosen for both reflectors without causing blockage by the location of either the array or the subreflector.

7.2.4 The DRA System

The reflector types of antennas have been successful in generating multiple shaped beams for regional coverage. These systems are primarily passive designs consisting of low-loss beamforming networks that feed all or part of the feed array. Individual channels are amplified and combined prior to beamforming. Because a beamforming network is designed for minimum loss, a compromise may be required in mass and volume. As the number of required beams increases to accommodate the higher capacity in a satellite system, the angular separation between beams decreases, and the reflector and feed array sizes increase accordingly. The attempt to push this technology further to increase the level of frequency reuse appears to be limited by the system mass and volume constraints.

The concept of eliminating the use of a reflector and using a directly radiating array has been investigated for satellite coverage [17–20]. When a direct radiating array is used, it must have an effective aperture comparable to that of the reflector in order to achieve the same spatial isolation between the cofrequency beams. If a phased array is fed with a low-loss analog beamforming network, its overall weight will make it less attractive as an alternative to the reflector system. However, if a DBF array is used, overall weight, efficiency, and reliability can be improved and the array becomes competitive with or even surpassing the reflector system, as illustrated by the examples given later. The array will incorporate power amplifiers directly before the radiating elements. The DBF array becomes especially attractive for generating a significant number of beams. With today's developing MMIC technology, a DBF antenna can be cost-effective and attractive for use in satellite systems.

7.2.5 System Characteristics

As discussed earlier, the signal at a feed element of the SRFF antenna system can be represented in terms of a beam. Therefore, the SRFF antenna system is suitable for beam-space processing. In the SRFF system, signals are focused by the reflector to a single element or a cluster of elements to form a beam. Multiple beams, which may not necessarily be orthogonal beams, can be generated by enabling multiple elements, but in general only a small fraction of the array elements (or beams) are active at any one time. Because the signals from the array elements are presented in terms of beams, it is natural for the beamforming process to be carried out in beam space even though the beams are not orthogonal.

The SRHT, DRAF, and DRA systems are suitable for element-space processing. In both the SRHT and DRAF systems, the signals arriving at the receivers are processed by two stages of Fourier transform. Therefore, the signal amplitude distribution across the transceivers is approximately uniform. In the DRA case, signals arrive at the array as plane waves with uniform amplitudes and linear phase change across the array. In any case, all the array elements simultaneously receive signals. Therefore, it is logical to employ element-space processing. That is, a desired beam can be formed by weighting and combining a fraction or all of the outputs from the array elements.

The implementation of DBF requires that each antenna element in the array has its own transceiver. In the SRFF system depicted in Figure 7.4, each transceiver is connected to one of the antenna elements. In the SRHT system shown in Figure 7.6, there are N independent transceivers, each of which is connected to one of the outputs of the Butler matrix. In both the DRAF and DRA cases, individual transceivers are connected to the antenna elements of the array. Failure of one or more transceivers has different consequences for the performance of these systems. In the SRFF system, energy is focused by the reflector on individual elements. Thus, failure of an element will lead to a rapid degradation in beam coverage. Therefore, a high redundancy for the transceivers is required. On the other hand, in the other three systems, energy is evenly distributed over the transceivers. Failure of a small fraction of the transceivers will only lead to a slow degradation in terms of the combined output power, sidelobe level, pointing accuracy, and beam isolation. If the combined output power is P_{out}, and one receiver failure decreases it to P'_{out}, then the rate of power decrease is given as [21]

$$\frac{P'_{\text{out}}}{P_{\text{out}}} = \left(\frac{N-1}{N}\right)^2 \tag{7.8}$$

which is considered a graceful degradation. Therefore, the redundancy requirement can be considerably lower.

All three types of reflector antennas are structurally simple, but they must be offset to avoid feed blockage. The offset design, however, destroys the rotational symmetry of the reflector's surface and limits the scan range. The destruction of the rotational symmetry reduces the possibility of frequency reuse based on polarization. The limited scanning ability does not present significant problems to a GEO system, since the requirement for scanning is not critical. In fact, the small scan angles from a geosynchronous orbit ($\pm 8°$ at most) simplify the design of the antenna, since they permit the array elements to be spaced far apart (e.g., $> \lambda/2$), allowing for large radiating elements and reducing problems of mutual coupling. Grating lobe effects due to large element spacing can be reduced by ensuring that the grating lobes from the feed array miss the reflector or, in the DRA case, miss the entire area to be covered. For a LEO system, a large scan range ($\gg 10°$) is required for the coverage of a small area. In this sense, the application of DBF using reflector types of antennas in a LEO system may not be feasible. In contrast, the ability of a DRA system to scan a wide angular range is excellent compared to that of reflector antennas. Furthermore, in low orbits, the requirements for high antenna gain are not as critical as in the GEO case. Therefore, the use of a DRA system in a LEO system is feasible and may be the only alternative for implementing DBF in LEO systems. Grating lobes must not be allowed in the LEO case, since they will certainly become sources of interference to users in other region. This can be ensured by making the antenna element spacing sufficiently small.

The reflector-type antennas, in general, have a relatively high directivity, which is proportional to the physical size of the reflector. To achieve the equivalent directivity at least to the first order, a direct radiating array must have an aperture equivalent to that of a reflector, which implies that the number of elements must be increased or the size of the elements must be relatively large to form an array with the required aperture. The active components that are required for constructing the array add the weight and complexity to the antenna system. However, the real physical aperture size of an array may be smaller than that of a reflector to have the same gain, because a direct radiating array has a higher aperture efficiency than the reflector system. That is, for the same physical aperture size, an array has a higher gain than a reflector. The low aperture efficiency of a reflector system is caused by loss due to spillover and aperture phase error, which increases with increasing scan angle.

7.3 SYSTEM EXAMPLES

Specific examples are given in this section to provide a quantitative illustration of the advantages that can be achieved using digital beamforming onboard a satellite system.

7.3.1 MSAT Systems

L-Band System

In an MSAT communications network, the mobile terminal employs a low-gain antenna of 3- to 6-dB gain (omnidirectional) or 10- to 14-dB gain (for steered antenna). Those low gain antennas collect the direct line-of-sight signal with its specular ground reflection components in the vicinity of the mobile terminal and ts multipath reflection components from hills, mountains, trees, buildings, and other large objects. The reflected components of the direct signal may add up destructively and result in signal fading. Therefore, an MSAT system is designed to compensate for the degradation due to propagation effect. MSAT services at L-band are allocated a narrow bandwidth of only 29 MHz. Thus, the system must be designed to efficiently utilize the frequency spectrum. The uplink frequency band for mobile terminals is 1.6316 to 1.6605 GHz and the downlink frequency band is 1.53 to 1.559 GHz. The satellite is capable of transmitting multichannel signals operated in continuous and intermittent modes (voice activation, data burst) to all service areas. The primary modulation format for single-channel and multichannel transmission of digital voice and data is trellis-coded modulation (TCM) with 8-ary PSK. The bit rate is 6200 bps. The minimum receive C/N_0 for the L-band links is normally 52 dB-Hz. The access method for both the uplink and downlink is FDMA. The channel spacing is 7500 Hz. Since the total available bandwidth is 29 MHz, the number of available channels is 3866. To apply frequency reuse the L-band transmit and receive antenna must spatially provide at least 18-dB cochannel isolation in the forward link case and 22-dB in the return link case.

In order to reuse the frequency spectrum, it is necessary to use a multiple beam satellite antenna for the coverage of Canada and the United States. Present analog beamforming technology permits frequency reuse by beams that are separated by more than two beamwidths to provide an adequate carrier-to-interference ratio. For the four-beam system shown in Figure 7.9, the two coastal beams can use the same frequency, and thus the frequency-reuse capacity can be increased by a factor of $\frac{4}{3}$.

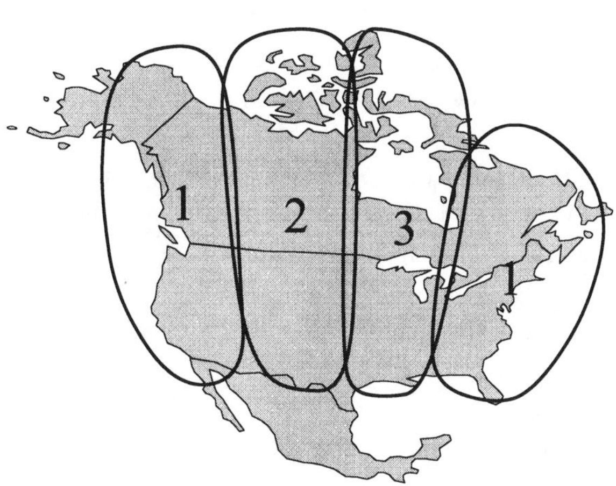

Figure 7.9 Four-beam coverage by analog beamforming in the L-band MSAT system.

Because MSAT services at L-band are allocated a narrow bandwidth of only 29 MHz, the system must be designed to efficiently utilize the frequency spectrum. DBF can be used to improve the frequency-reuse factor in that a relatively large number of beams can be formed and interference cancellation can be carried out to provide adequate carrier-to-interference ratio. This is in contrast to the analog beamforming technology, which only permits frequency reuse by beams that are separated by more than a beamwidth to provide an adequate carrier-to-interference ratio. For the four-beam system shown in Figure 7.9, the frequency-reuse capacity is $\frac{4}{3}$. On the other hand, for the case in which 12 beams are formed with DBF (Figure 7.10), the frequency reuse capacity can be increased by a factor of $\frac{12}{4}$ with respect to the single-beam scenario.

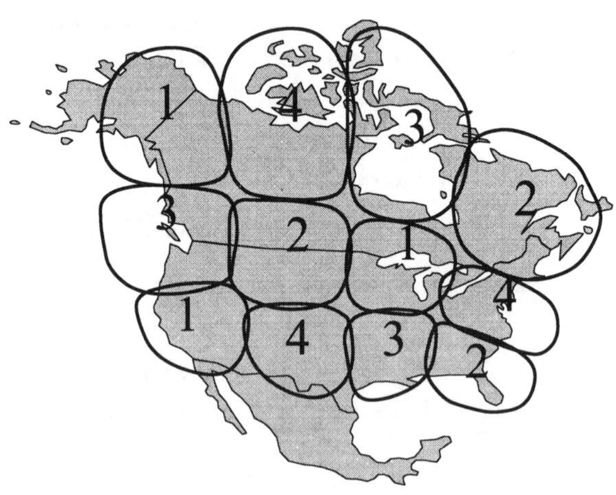

Figure 7.10 Twelve-beam coverage using DBF in the L-band MSAT system.

The SRHT configuration can be used in order to make use of the antennas in the existing MSAT system, where the number of feed elements for the L-band antenna is 23. Separated transmit and receive antenna systems are considered. The SRHT configuration requires the additional use of a hybrid transformer. As discussed previously, the SRHT increases reliability by distributing the power evenly to or from the antenna feed elements. In addition, it allows a greater flexibility for beamforming processing.

In the case of multiple users per beam, the access method for both the uplink and downlink is logically MCPC-FDMA. For each carrier or frequency subband within the available frequency band, multiple beams can be formed to achieve frequency reuse. Since a single beam provides multiple channels to users, communications signals are distributed within the HPBW area. The minimum cochannel beam separation is therefore determined by the requirement that in the forward link case, the -21-dB (-25-dB in the return link case) contour of the wanted beam should be outside of the -3-dB contour of the other cochannel beams. Orthogonal beams are not suitable, since they cannot provide low sidelobes and their separation is not sufficiently large in this case. The number of available beams per frequency subband depends on the beamwidth and the total required coverage area. Since the cochannel interference is required to be less than -18 dB, the cochannel beams must be separated by about two beamwidths. Since the message channel spacing is designed to be 7.5 kHz, the number of message channels per frequency subband is $\frac{29\ \text{MHz}/N_s}{7.5\ \text{kHz}} = 3866/N_s$, where N_s denotes the number of frequency subbands. Assuming 10% of the channels are reserved for call request, the number of available message channels per frequency channel is $0.9 \times 3866/N_s = 3480/N_s$. The total number of message channels available to users is $N_f \times N_s \times 3480/N_s = 3480N_f$. In the case shown in Figure 7.10, where $N_s = 4$ and $N_f = 3$, the total number of channels is 10,440.

The EIRP required by a mobile terminal determines their operation feasibility. The EIRP is given by

$$\text{EIRP} = \frac{C}{N_0}\frac{L_p L_0 N_0}{G_s} \tag{7.9}$$

where

$\frac{C}{N_0}$ = carrier-to-noise ratio;

L_p = propagation loss;

L_0 = other losses;

N_0 = noise power density;

G_s = gain of the satellite antenna.

The carrier-to-noise ratio is required by the design specification to be 52 dB-Hz. The propagation loss is 188.5 dB at the uplink frequency, assuming the geostationary orbit. Other losses include beam mispointing and atmospheric loss and are estimated to be 3 dB. The noise power density of the satellite system is estimated to be -202.9 dBW/Hz, assuming that the receiver noise temperature is 100K and the antenna noise temperature is 290K. The antenna gain is assumed to be 30 dB. Therefore, the required EIRP is 10.6 dBW. For the case of multiple users per beam, additional 3-dB power is required for mobiles at the edge of the beam coverage area.

Ka-Band System

The Ka-band system has been proposed to provide personal communication services including both voice and data. The uplink frequency band for mobile terminals is 29 to 30 GHz and the downlink frequency band is 19.2 to 20.2 GHz. The satellite is capable of transmitting multichannel signals operated in continuous and intermittent modes (voice activation, data burst) to all service areas. A 60-MHz frequency slot (54 MHz usable) will be available for the Ka-band uplink for mobile users (voice). Vocoded speech at 2400 bps and minimum shift-keying (MSK) modulation will be used for uplink. The modulation method used for downlink will be BPSK. The bit rate is 2400 bps for uplink and 43 kbps for downlink. The minimum receive C/N_0 for the Ka-band links is normally 45 dB-Hz. The access method for the uplink is tentatively FDMA and for the downlink TDMA. The channel spacing is 10 kHz for uplink and 150 kHz for downlink.

The EIRP required by a Ka-band mobile terminal can be computed using (7.9). The carrier-to-noise ratio is required by the design specification to be 45 dB-Hz. The propagation loss is 214.5 dB at the uplink frequency, assuming the geostationary orbit. The other losses are estimated to be 3 dB. The noise figure of the satellite system is estimated to be -202.9 dBW/Hz, assuming that the receiver noise temperature is 100K and the antenna noise temperature is 290K. The satellite antenna gain is assumed to be 40 dB. Therefore, the required EIRP for a mobile terminal is 19.6 dBW.

The DRAF system has been proposed as an option. The array consists of 343 elements, which are organized into a hexagonal arrangement, as shown in Figure 7.11. The distance between the centers of all the contiguous elements is two wavelengths (2λ). Accordingly, the diameter of the phased array is 40λ and

$$\mathrm{HPBW}_{\mathrm{array}} = \frac{1.02\lambda}{40\lambda} = 0.0255\ rad = 1.46^\circ \tag{7.10}$$

As a result, the HPBW of the DRAF system is

$$\mathrm{HPBW}_{\mathrm{system}} = \frac{1.46^\circ}{\mathcal{M}} \tag{7.11}$$

The value of $\mathcal{M}$ is chosen such that the size of the primary reflector produces the required gain. For $\mathcal{M} = 2.5$, which corresponds to a 1-m primary reflector, $\mathrm{HPBW}_{\mathrm{system}} = 0.58^\circ$.

The DRA has also been considered here as another option. In order to retain the same aperture gain as in the DRAF case, larger size antenna elements and/or more elements should be used. To minimize the hardware complexity, larger size antenna elements would be more preferable, since weight increase is mostly due to active components.

The number of degrees of freedom of the array is 343. Ideally, all of the degrees of freedom should be exploited for beamforming. However, this presents a huge computational problem. In order to minimize the computations, the use of subarrays

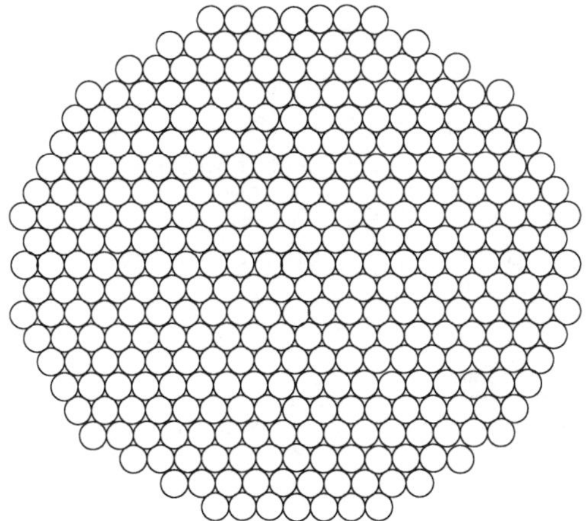

Figure 7.11 Antenna array configuration for the Ka-band system.

s suggested to. One option is to group three elements as a subarray, resulting in a 49-subarray array. Thus, the 49 degrees of freedom can be used in beamforming. It should be noted that the number of array elements can be different from the example given here. The actual value depends on the system requirements. Since the antenna array consists of a large number of elements, its performance degrades very slowly as elements fail. Therefore, minimal redundancy is desired to reduce the mass of the system.

For the case of single user per beam (SUPB), the access method for the uplink is the SCPC demanded-assigned multiple access (CPC-DAMA). For each carrier, multiple beams can be formed to achieve frequency reuse.

Since a single beam provides a single channel for one user, nulls can be steered towards the cochannel interfering beams. The minimum cochannel beam separation is therefore determined by how a null can be formed close to the main beam. The minimum cochannel beam separation can be less than one beamwidth. We assume one beamwidth for the purpose of illustration. The maximum number of beams per frequency channel (carrier) would be the maximum number of array elements or the number of one-beamwidth beams that are needed to cover the coverage area, whichever is less. In this case, it is 49. For uplink, since the message channel spacing is designed to be 10 kHz, the number of message channels per frequency channel is $\frac{54\text{MHz}}{10\text{kHz}}$ =5,400. Assuming 10% of the channels are reserved for call request, the number of available message channels is 0.9×5,400 = 4,866. The total number of message channels available to users is 49× 4,866 = 238,140. For downlink where multi-frequency (MF)-TDM is used, the number of channels per time frame is $49 \times 0.9 \times 54MHz/150kHz$ =15,876.

It should be pointed out that the calculation of the number of channels is based on the assumption of equal traffic in all beams. If the assumption does not hold, which would be a likely case, the number of channels may be decreased accordingly. Furthermore, the calculation is only based on the beamforming consideration. In an actual implementation, the number of channels may be relatively lower because it will certainly be limited by other factors, such as the capacity of the onboard switching circuit.

7.3.2 A VSAT System

The example of a very-small aperture terminal (VSAT) given here is a geostationary regional land mobile system [5]. The land mobile link is characterized by a particularly difficult shadowing and multipath environment, which leads to deep fades. A margin of 5 dB is assumed, comprising 2 dB for multipath and 3 dB for shadowing. Correlation in the shadowing fade for the mobile uplink and downlink may be exploited in the DBF system in controlling downlink channel power.

Specific assumptions used in the design include:

1. Regional service area is $10° \times 6°$.
2. Private networks are supported, each typically comprising a VSAT fixed terminal and a group of mobiles.
3. There are 4200 duplex voice channels (4.8-kbps, 3/4 coded quadrature phas-shift keying (QPSK), 5-kHz spacing, C/N_0 requirement of 41 dB-Hz).
4. Ku-band is used for the feeder link and L-band for the mobile link.
5. VSAT terminals have an EIRP level of 42.1 dBW and a 14-dB/K gain-to-noise-temperature ratio (G/T) (including static power control).
6. Mobile terminals have an EIRP level ranging from 11.8 to 15.8 dB and G/T of -12 dB/K (no power control).
7. Spatial frequency reuse is required.
8. The array has a size of 4m in diameter and 19 elements, resulting in a peak directivity of 34.2 dB and the maximum scan loss of 1 dB.

There are a number of considerations for the forward link budgets used to establish onboard power requirements. First, dynamic fades on the feeder uplink are not compensated for at the VSATs, so a 2-dB variation exists in the channel level as received at the satellite. In the case of analog beamforming, the downlink channel power level is determined based on the requirements for the weakest channels. On the other hand, this level variation can be compensated for in the case of digital beamforming. Second, the mobile link EOC gain in the analog approach is about 3 dB lower than that in the digital case, where a peak gain is available for the agile beams. Finally, in the analog approach, the downlink power level is set based on the weakest mobile link, whereas in the digital approach, level control can be used to compensate not only for static link variations but also for dynamic variations where there is a correlation between uplink and downlink frequencies, in particular mobile gain variations and shadowing fades. The actual power saving depends on the probability distribution in link performance and will be degraded by the response time of the loop. The estimated saving is about 4.5 dB when using a digital beamforming payload.

For the return link, there is a large variation in received power due to variations in mobile EIRP, receive gain, and fade depth. The total variation is estimated to be 12 dB. In the analog approach, the feeder downlink power must be based on the worst-case channel. In contrast, the digital approach allows channel power levels to be equalized, thereby reducing the required feeder link power. Approximate mass and power budgets for the analog and digital payload concepts are given in Table 7.1. The budgets show that in the digital case, there is a significant saving in high-power amplifier (HPA) primary power requirements for both the forward and the return links, which more than offsets the additional processor mass and power. The comparison between the systems is, however, critically dependent on a number of parameters. The advantage resulting from power level control is clearly dependent on variations in link performance between channels and hence on the fade margin and gain variation. For a given set of assumptions, the use of digital beamforming will reduce HPA power by a given factor such that the savings in absolute terms are larger for a large satellite, which is dominated by RF power requirements.

Six beams are required to cover the assumed area for the analog payload (HPBW of 3.3°). The frequency-reuse potential is limited by several factors. Due to the interference between cofrequency beams, the minimum spacing between beams based on main lobe interference constraints is required to be 1.6 HPBW. Isolation on the mobile uplink is degraded by differential link performance (estimated above 12 dB). The worst case is when the wanted signal is at its lowest level and the cochannel interference is at its highest level. Nonuniform traffic distribution also limits frequency reuse-potential. Given these factors, there is little scope for exploiting frequency reuse in this particular analog system. The required bandwidth for the 4200 channels is therefore 21 MHz.

The digital beamforming approach, however, offers significant advantages in terms of frequency reuse. Because there is a limited number of localized users per beam, the main lobe constraint allows a closer spacing between cofrequency beams, which is estimated to be 1.05 HPBW. Adaptive beamforming can provide

Table 7.1
Comparison of Payload Mass and Power Budgets.

	Analog Payload	*Digital Payload*
Forward HPA RF power	580.0W	73.0W
Forward HPA primary power	2000.0W	252.0W
Return HPA RF power	304.3W	48.2W
Return HPA primary power	1383.0W	219.0W
Feeder Rx/Tx system mass	60.0 kg	20.5 kg
Feeder Rx/Tx system power	37.0W	37.0W
Mobile antenna mass	36.0 kg	36.0 kg
Mobile Rx/Tx mass	90 kg	30 kg
Mobile Rx/Tx power	46.0W	46.0W
Processor mass	12.0 kg	31.5 kg
Processor power	41.0W	142.0W
Total payload mass	198.0 kg	118.0 kg
Total payload power	3507W	696W

a high level of isolation between cochannel beams by steering nulls towards the interferers. Adaptive signal processing may further improve the performance for the mobile uplink, especially when adverse differential link performance occurs with the interferer being stronger than the wanted signal. Adaptive control as applied to the uplink can also be used on downlink, but with some loss of performance. Given these factors, it is estimated that a spatial frequency-reuse factor of 3 can be realistically achieved given the beam size and service zone subject to a reasonable traffic uniformity. That is, the capacity provided by the digital payload is three times of that by the analog payload.

7.3.3 MAGSS-14

The European Space Agency (ESA) has proposed a MEO system concept called MAGSS-14 [8]. The network constellation consists of 14 satellites in circular orbits at 10,355 km. The minimum elevation angle achieved from the users to a satellite is 28.5° around the entire globe. This system is intended to offer land mobile personal communications to a variety of terminals, such as handheld, portable, and vehicle phone sets, using the recently allocated frequencies of 1.6138 to 1.6265 GHz and 2.4835 to 2.5000 GHz. The satellite has been sized to provide the equivalent of 1000 voice circuits to handheld terminals anywhere on the visible Earth (i.e., over all land masses and oceans from which the satellite is seen with more than 28.5° of elevation angle). The antenna radiation pattern and gain has to be adapted to the requirement for user mobility and expected degree of cooperation in pointing towards the satellite. The antenna profile and size is crucial for the terminal integration in a vehicle, suitcase, or a handheld set.

In the forward link case, because mobile users have limited transmitting power, it is required that the receive antenna onboard the satellite has a relatively high gain. If analog payload is used, the high gain requirement is met using spot beams to serve the coverage areas. The gain of the receive antenna has been chosen to be approximately 24.5 dBi (at the edge of the coverage) to obtain the required channel capacity with an onboard power of 1,500W to 1,600W. The gain also influences the number of beams that have to be generated to cover a area at a 30° minimum elevation angle. For this particular gain level, the number of beams has been determined to be 37, which is the result of the good coverage offered by the high minimum elevation angle of the constellation and the nadir pointing mode of the spacecraft. The RF power has been calculated for the worst case, where users are in the beams with the lowest peak gain at the edge of the coverage area. For users closer to the center of the coverage (i.e., near the subsatellite point) there is a power advantage resulting from the lower path loss and higher peak antenna gain. The actual benefit in total RF power requirement is directly dependent on the distribution of the users within the satellite coverage.

The onboard RF power can be decreased if higher minimum transmit gain is chosen. This, of course, will increase the number of beams, which is indeed a situation advantageous for the use of a digital beamforming system. The coverage is achieved by a large number of agile overlapped beams. Beamforming is carried out digitally at narrowband. Total control of element signals enables beam reconfiguration in the case of element chain failure or misalignment. Accurate user location for beam selection and/or pointing is required. Digital beamforming allows for implementation of a variety of direction finding algorithms.

The mobile antenna is a direct radiating array and comprises 61 elements. Digital beamforming is carried out to generate 91 beams, each of at least 26 dBi gain, to cover the entire satellite service area.

The total mass of the digital payload is 156 kg with a power consumption level of 1,300W. Its analog counterpart is 220 kg with a power consumption level of 1600W. The lower power consumption level by the digital payload is the result of the higher edge-of-beam gain selected for the payload. This is possible because of the much higher number of beams generated, achieving the same coverage with the same antenna, but with a higher degree of overlap.

The main advantages of using digital beamforming in this particular system are:

1. High RF power efficiency due to near-peak antenna gain;
2. Maximized routing flexibility due to very high granularity of only eight circuits for TDMA and six for FDMA;
3. Best frequency-reuse capability;
4. Compact feeder link (1.9 MHz for narrowband (NB)-TDMA and 2.5 MHz for FDMA, each way, without polarization reuse);
5. Peak satellite G/T for service to handheld terminals.

7.3.4 Other Systems

In a study given in [4], it was shown that a DBF system with a total equivalent mass (TEM) of 172.5 kg provided twice the number of channels as the analog Artemis system with a TEM of 184.2 kg. The TEM includes both the required payload mass and power under the conversion assumption that most spacecraft are designed to provide flexibility between payload mass and power with a typical equivalency of 0.1 kg/W. In the same study, a DBF system was compared to the INMARSAT-3 payload. While the DBF system would provide almost four times the capacity as the Inmarsat-3, its TEM is only about 60% of that of the Inmarsat-3.

References

[1] A. D. Craig et al., "Study on digital beamforming networks," Technical Report TP 8721, British Aerospace Ltd., Stevenage, England, July 1990.

[2] M. Barrett, "Digital beamforming network technologies for satellite communications," in *ESA Workshop on Advanced Beamforming Networks for Space Applications*, Noordjwijk, Netherlands, 1991.

[3] F. J. Lake and R. P. Curnow, "Active interference suppression and onboard source location within a digital beamforming payload," in *ESA Workshop on Advanced Beamforming Networks for Space Applications*, Noordjwijk, Netherlands, 1991.

[4] A. D. Craig et al., "A digital beamforming payload concept for advanced mobile missions," in *ESA Workshop on Advanced Beamforming Networks for Space Applications*, Noordjwijk, Netherlands, 1991.

[5] A. D. Craig, C. K. Leong, and A. Wishart, "Digital signal processing in communications satellite payload," *Electron. Comm. Eng. J.*, vol. 4, June 1992.

[6] H. G. Gockler and H. Eyssele, "A digital FDM-demultiplexer for beamforming environment," *Space Comm.*, vol. 10, pp. 197–205, 1992.

[7] T. Lo, C. Laperle, Y. Shen, M. Zhang, and J. Litva, "Digital beamforming for mobile satellite communications—Technology assessment and design study," Technical Report CRL Report no. 284, Communications Research Laboratory, McMaster University, Hamilton, Ont., Dec. 1993.

[8] P. Rastrilla, I. Stojkovic, J. Benedicto, and P. Rinous, "Medium altitude payload concepts for personal satellite communications," in *AIAA 15th Int. Comm. Satellite Systems Conf.*, San Diego, pp. 1186–1193, Feb. 1994.

[9] T. Gebauer and H. G. Gockler, "Channel-individual adaptive beamforming for mobile satellite communications," *IEEE J. Select. Areas Comm.*, vol. 13, pp. 439–448, Feb. 1995.

[10] L. J. Ricardi, "Multiple beam antenna," in A. W. Rudge, K. Milne, A. D. Olver, and P. Knight, eds., *The Handbook of Antenna Design*. Peter Peregrinus, London, 1986.

[11] A. W. Rudge and M. J. Whithers, "New technique for beam steering with fixed parabolic reflectors," *Proc. IEE*, vol. 11, pp. 857–863, July 1971.

[12] G. V. Borgiotti, "An antenna for limited scan in one plane: Design criteria and numerical simulation," *IEEE Trans. Antennas Propagat.*, vol. 25, pp. 232–242, March 1977.

[13] R. J. Mailloux, "Hybrid antennas," in A. W. Rudge, K. Milne, A. D. Olver, and P. Knight, eds., *The Handbook of Antenna Design*. Peter Peregrinus, London, 1986.

[14] M. H. Chen and G. N. Tsandoulas, "A dual reflector optical feed for wide-band phased array," *IEEE Trans. Antennas Propagat.*, vol. AP-22, pp. 541–545, July 1974.

[15] C. Dragone and M. J. Gans, "Image reflector arrangements to form a scanning beam using a small array," *Bell Syst. Tech. J.*, vol. 58, Feb. 1979.

[16] R. M. Sorbello, A. I. Zaghloul, B. S. Lee, S. Siddiqi, and B. D. Geller, "20-Ghz phased-arry-fed antennas utilizing distributed MMIC modules," *COMSAT Technical Review*, vol. 16, pp. 1184–1198, Fall 1986.

[17] W. Bornemann, P. Balling, and W. J. English, "Synthesis of spacecraft array antenna for INTELSAT frequency re-use multiple contour beams," *IEEE Trans. Antennas Propagat.*, vol. 33, pp. 1186–1193, Nov. 1985.

[18] A. I. Zaghloul, R. M. Sprbello, and F. T. Assal, "Development of phased arrays for reconfigurable satellite antennas," in *European Space Agency COST 213/KUL Phased Array Warkshop*, Leuven, Belgium, pp. 135–150, 1988.

[19] A. I. Zaghloul, Y. Hwang, R. M. Sorbello, and F. T. Assal, "Advances in multibeam communications satellite antennas," *Proc. IEEE*, vol. 78, pp. 1214–1232, Jan. 1990.

[20] R. M. Sorbello et al., "Independently steerable multiple-beam active phased-array antenna for future digital satellite applications," in *9th Int. Conf. on Digital Satellite Communications*, Copenhagen, pp. 206–210, 1992.

[21] S. Egami and M. Kawai, "An adaptive multiple beam system concept," *IEEE J. Sel. Areas Comm.*, vol. SAC-5, May 1987.

Chapter 8

Adaptive Beamforming in Mobile Communications

8.1 INTRODUCTION

The investigation into the use of adaptive arrays in communications began perhaps more than two decades ago [1]. The objective then was to develop receiving systems for acquiring desired signals in the presence of strong jamming, especially in military communications. Adaptive array systems have been developed for receiving TDMA satellite communications signals [1] and spread-spectrum communications signals [2, 3].

The use of antenna arrays in mobile radio systems to combat cochannel interference was first discussed by Yeh and Reudink in 1980 [4]. They showed that with a large number of antenna elements, it was possible to carry out frequency reuse to achieve a very high frequency spectrum efficiency. In an in-depth discussion [5], they observed that with a moderate number of space diversity branches, much higher spectrum efficiency was achievable, without a spectral penalty, compared to a non-diversity system. Using 24 diversity branches for the basestation and 20 branches for the mobile, 100% relative spectrum efficiency can be achieved. A similar concept to that of Yeh and Reudink was independently suggested by Bogachev and Kiselev around the same time [6]. They derived the expression for the probability of the signal-to-interference ratio in the presence of Rayleigh fading. The capability of adaptive arrays and their potential to increase the spectrum efficiency in mobile communications has been gradually recognized by the industry. The analysis given in [7] shows that adaptive beamforming technology allows a reduction in basestation spacing by up to 40%, thereby increasing the reuse of radio channels by a factor of 2.8. In his classic paper [8], which may arguably be viewed as a milestone in this particular technical field, Winters presented an in-depth study of optimum signal combining for space diversity reception in cellular mobile radio systems. In the optimum-combining approach, the signals received by the antennas are so weighted and combined that the signal-to-interference-plus-noise ratio (SINR) of the output is maximized, thereby reducing both the effects of Rayleigh fading of the desired signal and the cochannel interference. It should be pointed out here again that optimum combining is also called adaptive beamforming, especially in the antenna community. We will use both terms interchangeably. The results that Winters obtained show that optimum combining is significantly better than maximum ratio combining even when the number of cochannel interferers is greater than the number of antenna elements. Furthermore, optimum combining increases the output SIR at the receiver by several decibels. Since then, the concept of using adaptive antennas in land mobile communications has been well received by the communications community, especially in recent years, when PCS has been in its development stage. Many technical papers have been published on adaptive beamforming technology for mobile communications and PCS. A few notable examples are [9–12].

8.2 BENEFITS OF USING ADAPTIVE ANTENNAS

We have mentioned in a previous chapter some general advantages of using digital beamforming. Here, we will discuss the specific benefits of using adaptive antennas in mobile communications. With the adaptive antenna technology, many of the system parameters that are considered constraints in a single-antenna system become parameters that a system designer has at his disposal to further optimize system performance. It is this aspect of adaptive antenna technology which is perhaps most attractive with regard to its inclusion in the next generation of PCS systems to be implemented. In particular, a mobile communications system or a PCS system may benefit from the use of adaptive antenna technology in the following aspects.

Coverage

Adaptive beamforming can increase the cell coverage range substantially through antenna gain and interference rejection. In particular, the coverage range can be improved by a factor of $M^{1/n}$ for noise-limited environments, where M denotes the number of antenna elements and n denotes the propagation loss exponent. For example, if $n = 4$ and $M = 16$, the range is approximately doubled with respect to the omnidirectional antenna case

Generally, there will be fewer sites required with adaptive antennas employed at base stations. A larger coverage area can be achieved with basestation antennas at a greater height above average terrain. However, this system trade-off between antenna height and coverage area can be eased by using the number of antenna elements as a design parameter. This leads to substantially eased siting requirements in many situations.

Capacity

Adaptive antenna technology provides the flexibility that, in conjunction with centralized dynamic channel assignment (DCA) strategy, allows a reuse factor of unity; that is, a single frequency can be used in all cells. Reuse planning therefore becomes a dynamic programming problem that is solving the problem of maximizing the number and quality of mobile links. With respect to CDMA systems, the reuse is code reuse rather than frequency reuse, but the fundamental concept is the same.

In general, adaptive beamforming can increase the number of available voice channels through directional communication links. The increase factor depends on the propagation environment, the number of antenna elements, and the amount of DCA allowed by the system. The point is that it is possible to have multiple mobiles on the same RF channel but different spatial channels at a particular cell site. Furthermore, frequency-reuse patterns can be substantially improved or reduced though intelligent use of adaptive antennas.

Transmission bit rate can be increased due to the improved SIR at the output of the adaptive beamformer. The minimum rate improvement would be that attainable with $10 \log M$ dB SIR improvement in noise-limited environments and depends on the modulation format. Adaptive antennas allow RF channels to be adjusted through link power control to meet the requirements of user-selectable data transfer rates. Low-rate users can use lower power links; high-rate users require higher power links.

Signal Quality

In noise-limited environments, minimum receiver thresholds are reduced by $10 \log M$ dB on average. In interference-limited environments, the additional improvement in tolerable SIR at a single element results from interference rejection afforded against directional interferers. The level of improvement depends on the distribution of cochannel users in neighboring cells. With centralized DCA, it is possible to improve system performance even further by providing optimization on a "global" system-wide basis.

Adaptive antennas can be considered as spatial equalizers and can provide substantial signal quality improvements through spatial signal processing. In fact, some implementations of adaptive antennas provide a spatial rake receiver capability to combine uplink multipath arrivals for improved output SIR. There is virtually no limit to the amount of delay spread tolerable on the uplink, since the signals are with probability >95% spatially distinguishable. The downlink is unchanged except for the statistical improvement achievable because of the reduced RF pollution.

Access Technology

Adaptive beamforming provides a dimension to the radio interface in addition to any or all temporal modulation techniques such as FDMA, TDMA, and CDMA. The benefits are multiplicative; that is, temporal gains are multiplied by spatial gains.

Equalization is inherent in uplink processing where paths from different angles of arrival are separated be using a particular adaptive beamforming technique. On the downlink, energy can be focused at the mobile so that long delay multipath components can be reduced substantially. This enables the system to combat ISI through spatial discrimination of "interfering" signals on both forward and reverse links. The system can handle a rapidly varying delay spread profile on both links as well as through appropriate signal processing at the base station on both transmit and receive.

Power Control

Power control requirements of the various modulation methods are somewhat eased through the inclusion of adaptive antenna technology; however, there are robustness benefits to be attained through coarse mobile unit and base station power control. This dynamic link budget control is normally a difficult process, but it can be made substantially easier through the use of angular information about the user signals provided by adaptive antennas.

Power control can be enforced to ensure maximum use of dynamic range. The objective is to keep all the users at a power level as low as possible under the constraint that the range of power levels seen at the base stations is within 10 to 20 dB.

Handover

In many cases, adaptive antenna technology provides mobile unit location information that can be used by the system to substantially improve handoffs in both the low and high tiers. With sufficiently accurate position estimates, prediction of velocities is possible which allows further improvements in handoff strategies.

Handover is performed by passing mobile unit tracking information from the base stations to the control center for resource allocation optimization. Deciding which cell to hand a mobile to is a much easier task when one knows where the unit is and how fast it is moving. Coupled with the good engineering practice of designing overlapping coverage areas, the result is "smart" handoff, which is neither "soft" nor "hard".

Basestation Transmit Power

Using adaptive beamforming, the maximum peak EIRP required per user on a particular channel is $10 \log M$ dB less than without adaptive beamforming. The average EIRP is similarly reduced. On a per element basis, $10 \log M$ less power is transmitted while the antenna array is still able to maintain the same power level at the mobile unit as in the case without the use of an adaptive antenna.

Portable Terminal Transmit Power

If an adaptive antenna is implemented in a system without changing other parameters (e.g. cell size), the transmission power levels required for portable terminals can be reduced on average by at least $10 \log M$ dB. The reduction in transmission power levels, which results from the increase in antenna gain at the basestation, relaxes the requirements on batteries. Other relevant issues include increased fade margin for improved signal quality (e.g., higher data rates, increased coverage area per cell to decrease deployment costs). With appropriate link management in the system, the power budgets for each of the radio links can be optimized for each particular link dynamically, a process made substantially easier through the use of mobile position estimates provided by adaptive antennas. The key point is that with adaptive antennas at cells, the transmit power levels from and to the mobile can be kept minimum to provide the requested service. The power levels will not exceed those that would otherwise be transmitted from an onmi or hard-sectored site.

8.3 PERFORMANCE IMPROVEMENT

In the previous section, we qualitatively discussed the impacts of adaptive antennas on land mobile communications systems. In the section, we will quantify the performance improvement brought about by the use of an adaptive antenna in a mobile communications system. The performance of a mobile communications system can be characterized by its bit error rate (BER). BER gives a good indication of the performance of a particular modulation scheme. However, it does not provide information about the types of errors. For example, it does not show incidents of bursty errors. In a fading mobile channel, it is likely that a transmitted signal will suffer deep fades, which can lead to outage or a complete loss of the signal. Probability of outage is another performance measure for a mobile communications system. An outage event is specified by a specific number of bit errors occurring in a given transmission

8.3.1 BER Performance with Fading

Let us assume the following:

1. Flat fading across the channel;
2. Independent fading between antenna elements;
3. A complex Gaussian distribution for the received signal.

In a mobile communications system, there are a relatively large number of cochannel interfering signals, compared to the number of antenna elements. An exact analysis of the performance is quite complicated, especially since each of these interfering signals has a random amplitude due to fading. To simplify the analysis, only the strongest interferers are individually considered. The rest is considered as a lumped interference that is uncorrelated between elements [8]. Furthermore, since the interfering signals has a Gaussian distribution, the sum of these signals also have a Gaussian distribution and can be considered as thermal noise. In the worst case, it is assumed that the adaptive antenna is not able to suppress the lumped interference. If only a single interferer is considered, the pdf of the SINR is given by [6]

$$p(\gamma_a) = \frac{e^{-\frac{\gamma_a}{\Gamma_e}} \left(\frac{\gamma_a}{\Gamma_e}\right)^{M-1} (1+\Gamma_T)}{\Gamma_e (M-2)!} \int_0^1 e^{-\frac{\gamma_a \Gamma_T}{\Gamma_e} t} (1-t)^{M-2} \, dt \tag{8.1}$$

where

γ_a = SINR at the output of the array;
Γ_e = SNR at each antenna element;
Γ_T = $M\Gamma_i$ = total interference-to-noise ratio;
Γ_i = interference-to-noise ratio (INR) of the ith interferer at each antenna elem

It follows that the corresponding cumulative probability is given by

$$P(\gamma_a) = \int_0^{\frac{\gamma_a}{\Gamma_e}} \frac{e^{-\zeta} \zeta^{M-1} (1+\Gamma_T)}{(M-2)!} \int_0^1 e^{-\zeta \Gamma_T t} (1-t)^{M-2} \, dt \, d\zeta \tag{8.2}$$

If there is no interference (i.e., $\Gamma_T = 0$), the system will behave as a maximum-ratio combiner and $P(\gamma_a)$ is given by

$$P(\gamma_a) = 1 - e^{-\frac{\gamma_a}{\Gamma_e}} \sum_{m=1}^{M} \frac{\left(\frac{\gamma_a}{\Gamma_e}\right)^{m-1}}{(m-1)!} \tag{8.3}$$

which is identical to (5.17). With a single strong interferer (i.e., $\Gamma_1 \Rightarrow \infty$), $P(\gamma_a)$ is given by

$$P(\gamma_a) = 1 - e^{-\frac{\gamma_a}{\Gamma_e}} \sum_{m=1}^{M-1} \frac{\left(\frac{\gamma_a}{\Gamma_e}\right)^{m-1}}{(m-1)!} \tag{8.4}$$

That is, adaptive beamforming with a strong interferer gives the same results as maximum-ratio combining without interference and with one less antenna element. The BER for coherent detection of PSK is given by

$$\text{BER} = \int_0^\infty p(\gamma_a)\,\text{erfc}\left(\sqrt{\Gamma\gamma_a}\right) d\gamma_a \tag{8.5}$$

In the case of a single interferer, the BER is given by [6]

$$\begin{aligned} \text{BER} \;=\; & \frac{(-1)^{M-1}(1+\Gamma_T)}{2\Gamma_T^{M-1}} \left\{ \frac{-\Gamma_T}{(1+\Gamma_T)} + \sqrt{\frac{\Gamma_e}{1+\Gamma_e}} - \frac{1}{(1+\Gamma_T)}\sqrt{\frac{\Gamma_e}{1+\Gamma_T+\Gamma_e}} \right. \\ & \left. - \sum_{m=1}^{M-2} (-\Gamma_T)^m \left[1 - \sqrt{\frac{\Gamma_e}{1+\Gamma_e}} \left(1 + \sum_{k=1}^{m} \frac{(2k-1)!!}{k!\,(2+2\Gamma_e)^k} \right) \right] \right\} \end{aligned} \tag{8.6}$$

where

$$(2k-1)!! = 1 \cdot 3 \cdot 5 \cdots (2k-1) \tag{8.7}$$

In essence, (8.6) exhibits the following relationship:

$$\text{BER} \propto \frac{1}{(M\Gamma_1)^{M-1}} \tag{8.8}$$

That is, the improvement in BER is mostly dependent on Γ_1 and M. Figure 8.1 [8] shows the BER versus the average received SINR for optimum combining with a single interferer. The results for $\Gamma_1 = 0$ corresponds to those for maximum ratio combining. One can observe that the improvement in terms of the received SINR required for a given value of BER is nearly independent of the SINR for BER $< 10^{-2}$.

When there is more than one interferer, it is very difficult to determine analytically the BER performance. Therefore, researchers and engineers rely almost exclusively on Monte Carlo simulation to determine the BER performance. For two interferers, the BER results are shown in Figure 8.2 [8]. The improvement obtained using optimum combining increases with the number of antenna elements. For example, for BER $= 10^{-3}$ and $M = 5$, optimum ombining requires 4.2 dB less than maximum-ratio combining.

8.3.2 Cochannel Interference Reduction

As we mentioned earlier, the use of adaptive antennas can reduce cochannel interference. In this subsection, we will discuss how much cochannel interference reduction

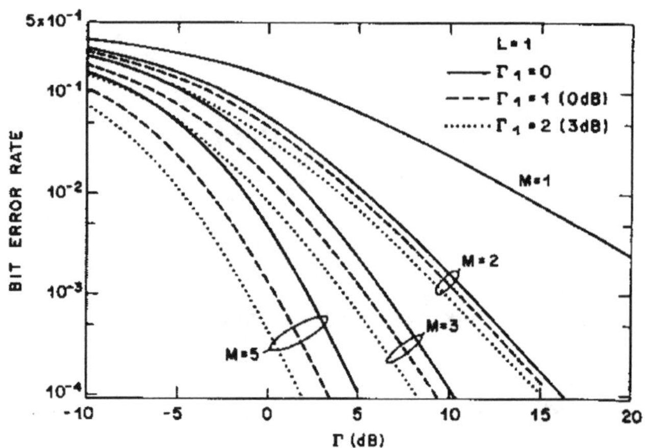

Figure 8.1 Average BER versus average SINR for optimum combining with a single interferer. (From [8] ©1984 IEEE.)

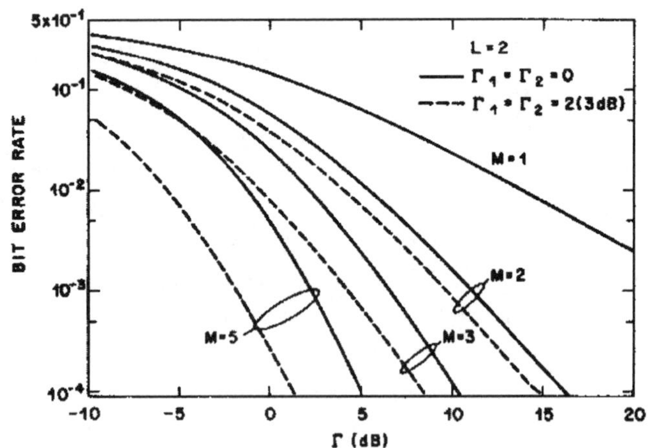

Figure 8.2 Average BER versus average SINR for optimum combining with a pair of interferers. (From [8] ©1984 IEEE.)

can be achieved. The reduction is quantified by comparing the outage probabilities when omnidirectional basestation antennas are used and when adaptive antennas are used. In the analysis given here, which largely follows that in [10], the following scenario is assumed:

1. A cellular network consists of hexagonal cells, with channel reuse every C cells, where C is the cluster size.
2. The basestation transmitters are located at the centers of the cells.
3. Users are uniformly distributed in each cell.
4. The blockage in each cell has a probability denoted by p_B.
5. The omnidirectional basestation antenna has an ideal beam pattern, giving a uniform circular coverage.
6. The adaptive basestation antenna is able to generate a number, m, of ideal beams with a beamwidth of $\frac{2\pi}{m}$ and a gain equal to that of the omnidirectional antenna.
7. Each adaptive beam will only carry the channels that are assigned to the mobiles within its coverage area.
8. The adaptive basestation antenna is able to track the mobiles;

The blockage probability p_B is the fraction of attempted calls that cannot be allocated a channel. If there are a Erlangs of traffic intensity, the actual traffic carried is equal to $a(1 - p_B)$ Erlangs. The outgoing channel usage efficiency can be defined as

$$\eta = \frac{a(1 - p_B)}{N} \tag{8.9}$$

where N is the total number of channels available per cell. Furthermore, cochannel interference is said to occur when the ratio of the desired signal power σ_d^2 to the interfering signal power σ_i^2 is less than some protection ratio p_r; that is,

$$\gamma_s = \frac{\sigma_d^2}{\sigma_i^2} \leq p_r \tag{8.10}$$

Since γ_s is a statistical quantity, it is necessary to introduce the concept of outage probability $p(\gamma_s \leq p_r)$, which is the probability of failing to obtain satisfactory reception in the presence of interference.

In a simple network where there are only two cochannel cells, based on the assumption of a equal blockage probability in all cells, the number of active channels in each cell is $N\eta$. In the case of omnidirectional basestation antennas, given that the desired mobile is already allocated a channel, the probability of this particular channel being in an interfering cell is the required outage probability. Thus, when the desired mobile is in the zone of cochannel interference, the outage probability is given by

$$p(\gamma_s \leq p_r) = \frac{N\eta}{N} = \eta \tag{8.11}$$

In the case of adaptive antennas, there are m beams in a cell and $\frac{N\eta}{m}$ channels per beam. If the desired mobile is in the zone of cochannel interference, it is always covered by at least one beam from the cochannel cell. Therefore, the outage probability is the probability that one of the channels provided by this particular beam is the corresponding active cochannel, which is the ratio of the number of channels per beam to the total number of channels; that is,

$$p(\gamma_s \leq p_r) = \frac{N\eta/m}{N} = \frac{\eta}{m} \tag{8.12}$$

The above analysis can be extended to the case where there are six cochannel cells in the network. In this case,

$$\gamma_s = \frac{\sigma_d^2}{\sigma_I^2} \tag{8.13}$$

where σ_I^2 is the total mean power resulting from all the cochannel interfering cells. In this case, six beams will be pointed to the desired mobile at any time, since m beams are formed in each cell. The outage probability within the region of cochannel interference is determined by considering the probability that the active cochannel is in each of these beams; that is,

$$p(\gamma_s \leq p_r) = \left(\frac{\eta}{m}\right)^6 \tag{8.14}$$

The above analysis is quite simplistic and optimistic in that it is based on a geometrical model. In practice, there is seldom a line-of-sight (LOS) path between the basestation and the mobile. Radio communications is obtained by means of diffraction and reflection of the transmitted signals. In order to obtain a more accurate evaluation, propagation effects should be included, in a statistical fashion, in the analysis. In general, there are two types of propagation phenomena. The first type is fading, which is due to the multipath nature of the received signal. Fading, which causes the signal level to fluctuate rapidly, usually follows a Rayleigh distribution. The pdf of the signal envelope due to fading is given by

$$p_f(s) = \frac{\pi s}{2\overline{s}^2} e^{-\frac{\pi s^2}{4\overline{s}^2}} \tag{8.15}$$

where $\overline{s} = E[s]$. The second type is shadowing, which is due to variation in the local terrain. Shadowing causes the average signal level to vary about the local mean. Shadowing follows a log-normal distribution. The pdf of the average signal envelope due to shadowing is given by

$$p_S(v) = \frac{1}{\sqrt{2\pi}\sigma_v} e^{-\frac{(v-\overline{v})^2}{2\sigma_v^2}} \tag{8.16}$$

where $v = 20\log \overline{s}$, $\overline{v} = E[v]$ and $\sigma_v^2 = \text{VAR}[v]$. The combined pdf can be written as

$$\begin{aligned} p(s) &= \int_{-\infty}^{\infty} p_f(s) p_S(v) dv && (8.17) \\ &= \frac{\sqrt{\pi}}{2\sqrt{2}\sigma_v} \int_{-\infty}^{\infty} \frac{s}{10^{v/20}} e^{-\frac{\pi s^2}{4\times 10^{v/10}}} e^{-\frac{(v-\overline{v})^2}{2\sigma_v^2}} dv && (8.18) \end{aligned}$$

Using (8.18), it can be shown that the outage probability in the case of one cochannel cell with omnidirectional basestation antennas is given by [13]

$$p(\gamma_s \le p_r) = \frac{1}{\sqrt{\pi}} \int_{-\infty}^{\infty} \frac{e^{-u^2}}{1 + 10^{(\varsigma - 2\sigma_s u)/10}} du \tag{8.19}$$

where $\varsigma = 20\log \frac{\overline{s}_d}{\sqrt{p_r s_i}}$ and $\sigma_s = \sigma_d = \sigma_i$. Based on the same arguments as in the derivation of (8.12), the outage probability is simply expressed as

$$p(\gamma_s \le p_r, m) = p(\gamma_s \le p_r)\frac{\eta}{m} \tag{8.20}$$

To evaluate the outage probability in the case where there are six cochannel cells in a network, the total interference has to be considered; that is,

$$s_I = \sum_{n=1}^{N_I} s_{i_n} \tag{8.21}$$

where s_{i_n} is the interference signal originating from the nth cochannel cell and N_I is the number of active interfering cochannel cells. In practice, N_I is a random process and has a pdf $p(N_I)$ given by [10]

$$p(N_I) = \begin{pmatrix} 6 \\ N_I \end{pmatrix} \left(\frac{\eta}{m}\right)^{N_I} (1 - \frac{\eta}{m})^{6-N_I} \tag{8.22}$$

which is also called the origination probability. Thus, the probability that there is cochannel interference and N active cochannel cells is given by

$$p(\gamma_s \le p_r, N_I) = p(\gamma_s \le p_r | N_I) p(N_I) \tag{8.23}$$

where $p(\gamma_s \leq p_r | N_I)$ is the conditional outage probability given N_I active cochannel cells. The total outage probability can be determined by taking into account all the possible values of N_I; that is,

$$p(\gamma_s \leq p_r) = \sum_{N_I} p(\gamma_s \leq p_r, N_I) = \sum_{N} p(\gamma_s \leq p_r | N_I) p(N_I) \tag{8.24}$$

The conditional outage probability is given by [14]

$$p(\gamma_s \leq p_r \mid N_I) = \frac{1}{2\pi} \int_{-\infty}^{\infty} dX \int_{-\infty}^{\infty} \left(1 - e^{-\frac{\pi}{4K^2(X,u)}}\right) e^{-\frac{X^2+u^2}{2}} du \tag{8.25}$$

where

$$20 \log K(X, u) = \varsigma + c \ln\left(\frac{4}{\pi N_I^2}\right) + \sigma X - \sigma_{NY} u - \frac{1}{4c}\left(\sigma_{NX}^2 - \sigma_{NY}^2\right) \tag{8.26}$$

$$c = \frac{10}{\ln 10} \tag{8.27}$$

$$\sigma_{NX}^2 = \sigma^2 + 4c^2 \ln\left(\frac{4}{\pi}\right) \tag{8.28}$$

and

$$\sigma_{NY}^2 = 4c^2 \ln \frac{e^{\frac{\sigma_{NX}^2}{4c^2}} + n - 1}{n} \tag{8.29}$$

The overall outage probability can be written as

$$P(\gamma_s \leq p_r) = \sum_{N} P(\gamma_s \leq p_r \mid N) \binom{6}{N_I} \left(\frac{\eta}{m}\right)^{N_I} (1 - \frac{\eta}{m})^{6-N_I} \tag{8.30}$$

The outage probability for $m = 1$ corresponds to the case of omnidirectional antennas. Figure 8.3 [10] shows the variation in the total outage rate as a function of ς. The fading and shadowing effects have been considered in the evaluation of the outage probability. Figure 8.4 [10] gives the relative spectrum efficiency for a given outage criterion (1%), where the outgoing channel efficiency is 0.7 and a log-normal shadowing standard deviation is 6 dB. It can be observed that the use of an adaptive antenna with eight tracking beams can increase the efficiency by as much as three times.

8.3.3 Improvement in CDMA Systems

We have shown that, in general, the use of adaptive antennas improves the performance of a mobile communications system in terms of BER and outage probability. In this section, we will discuss the performance enhancement brought about by the use of adaptive antennas in an interference-limited asynchronous CDMA system. In order to facilitate the following analysis, a number of assumptions have to be made:

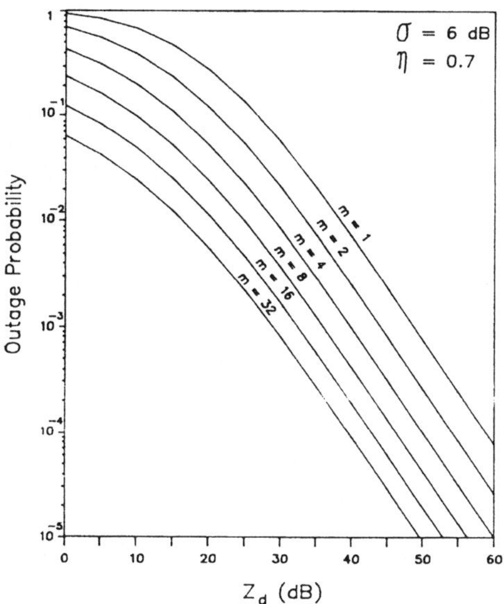

Figure 8.3 Outage probability with six cochannel cells. (From [10] ©1990 IEEE.) Note that Z_d for the x-axis label is equivalent to ς in the text.

1. The CDMA system operates in an additive white Gaussian noise (AWGN) channel.
2. Perfect power control is achievable.
3. Spreading factor is N_s.
4. There are K users uniformly distributed in each cell.
5. The pattern of a beam formed by the array does not vary in the elevation plane.
6. A beam formed by the array can be steered in any direction in the azimuth plane such that the desired mobile is always illuminated by the main beam.

If there is no interference from adjacent cells and an omnidirectional basestation antenna is used, the BER can be approximated by [15]

$$\text{BER} = Q\left(\sqrt{\frac{3N_s}{K-1}}\right) \tag{8.31}$$

where

$$Q(x) = \frac{1}{\sqrt{2\pi}} \int_x^{\infty} e^{-\frac{y^2}{2}} dy \tag{8.32}$$

On the reverse link, if a beam with a directional gain $G(\phi)$ is formed in the direction ϕ of the desired signal from a particular user, the signals from the other $K-1$ users

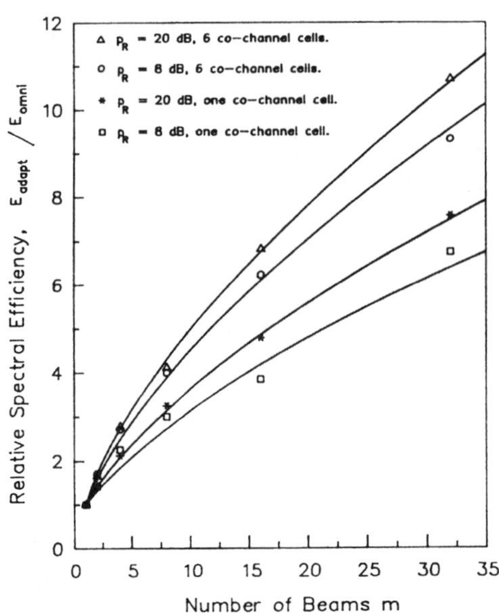

Figure 8.4 Relative spectral efficiency as a function of the number of beams. (From [10] ©1990 IEEE.)

are considered to be interference. The average total interference power σ_I^2 with respect to the signal of this particular mobile is given by

$$\sigma_I^2 = E\left[\sum_{k=1}^{K-1} G(\phi_k)\sigma_k^2\right] \tag{8.33}$$

where ϕ_k denotes the direction of the kth interfering mobile signal. With perfect power control, the signals from all the mobiles have the same power σ_c^2. Thus,

$$\sigma_I^2 = \sigma_c^2\, E\left[\sum_{k=1}^{K-1} G(\phi_k)\right] \tag{8.34}$$

By taking into account the pdf that describes the geographical distribution of the mobiles, σ_I^2 is found to be [16]

$$\sigma_I^2 = \frac{\sigma_c^2(K-1)}{D} \tag{8.35}$$

where D represents the array antenna's directivity given as

$$D = \frac{2\pi}{\int_0^{2\pi} G(\phi)d\phi} \tag{8.36}$$

It can be shown that the BER can be approximated by [16]

$$\text{BER} = Q\left(\sqrt{3N_s \times \text{SIR}}\right) \tag{8.37}$$

Since SIR= $\frac{\sigma_d^2}{\sigma_I^2}$, the BER is therefore given by

$$\text{BER} = Q\left(\sqrt{\frac{3N_s D}{K-1}}\right) \tag{8.38}$$

Based on the intuition that the additional interference from adjacent cells simply adds to the interference level, the average probability of error for a particular user in a multiple-channel-cell environment is given by [16]

$$\text{BER} = Q\left(\sqrt{\frac{3N_s D f_r}{K-1}}\right) \tag{8.39}$$

Figure 8.5 shows the BER performance versus the relative directivity of the array antenna, with $N_s = 64$ and $K = 63$. In the case of the omnidirectional antennas, $D = 0$ dB. It can be observed from the figure that considerable improvement in terms of BER can be obtained when an antenna array is used.

In order to evaluate the outage probability, it is necessary to model the interference using the appropriate statistics. Let us consider a particular case in which the unlink signal from a mobile user is subject to interference from both users in the same cell as this user and from users in adjacent cells. We can express

$$\rho_1 = \sum_{k_1=2}^{K} \varphi_{k_1} \tag{8.40}$$

as the interference-to-signal power ratio due to cocell users and

$$\rho_2 = \sum_{l=1}^{L} \sum_{k_l=1}^{K} \varphi_{k_l} \beta_{k_l}^2 \tag{8.41}$$

as the interference-to-signal power ratio due to adjacent-cell users. In both the expressions, φ is a random variable with a mean of ν, representing both the voice activity and the interfering activity of a particular user. The voice activity is the probability that a user is talking, whereas the interfering activity is the probability that an interfering user fall into the main beam that is assigned to the desired user. β^2 denotes the attenuation factor due to propagation. It should be noted that in the case of a cocell, β^2 is equal to one based on the assumption of perfect power control. ρ_1 is also a random variable, which has a binomial distribution with

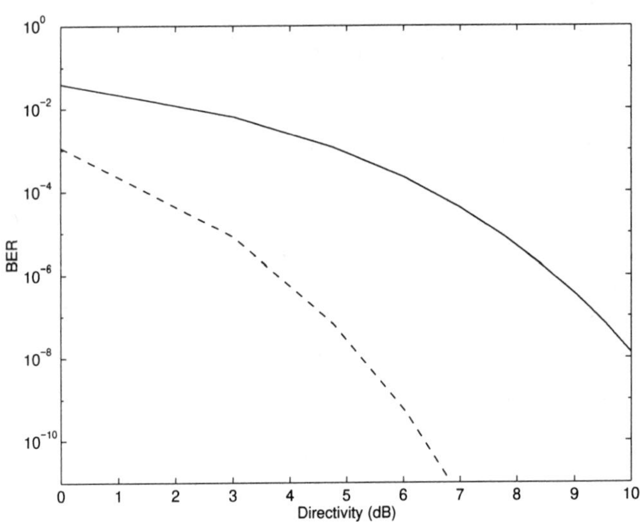

Figure 8.5 BER performance as a function of relative directivity; $D = 0$ dB for an omnidirectional basestation antenna. Dashed line: $f_r = 3$; solid line: $f_r = 1$

parameters $(K-1, \nu)$. For a large value of K, the random variable ρ_2 can be approximated by a Gaussian random variable with a mean of $\mu_i K$ and a variance of $\sigma_i K$ that depend on ν, the degree of shadowing, and the number of interfering cells L. The outage probability is

$$p(\gamma_s \leq p_r) = p\left(\rho_1 + \rho_2 > \frac{N_s}{p_r} - \frac{\text{SNR}}{M}\right) \tag{8.42}$$

where γ_s denotes the SINR, which is given by

$$\gamma_s = \frac{N_s}{\rho_1 + \rho_2 + \dfrac{\text{SNR}}{M}} \tag{8.43}$$

Based on the analysis given in [17], which is after [18], the outage probability can be expressed as

$$p(\gamma_s \leq p_r) = \sum_{k=0}^{K-1} \begin{pmatrix} K-1 \\ k \end{pmatrix} \nu^k (1-\nu)^{K-1-k} Q\left(\frac{\Delta - k - \mu_i K}{\sqrt{K}\sigma_i}\right) \tag{8.44}$$

where $\Delta = \frac{N_s}{p_r} - \frac{\text{SNR}}{M}$. The improvement in outage probability can be seen from the fact that as the number of antenna element M increases, the magnitude of $Q(\cdot)$

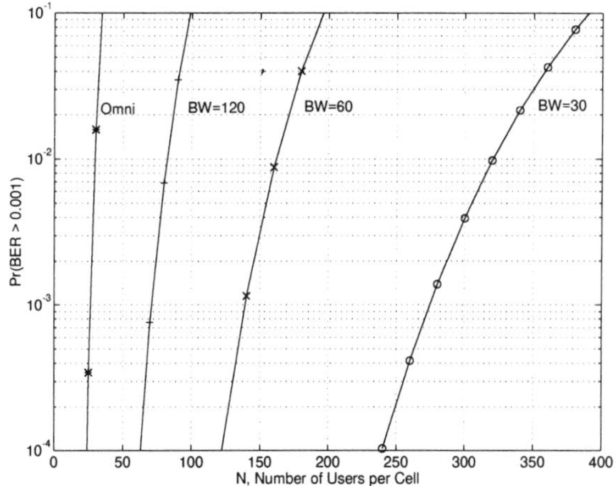

Figure 8.6 Uplink outage probability as a function of beamwidth and cell loading. (From [17] ©1994 IEEE.)

decreases, and hence the outage probability decreases. An example of the outage probability is shown in Figure 8.6.

In the downlink case, the approximation to the interference statistical model is not as readily found as in the uplink case. Thus, simulations were used to evaluate the downlink outage probability [17]. A set of simulation results are depicted in Figure 8.7, which show considerable increase in capacity when an antenna array is used.

8.4 ADAPTIVE BEAMFORMING FOR UPLINK

In this section, we will focus our discussion on the use of adaptive antennas in the uplink or reverse link (i.e., mobile to basestation) case. For years, much research effort has been devoted to the studies of adaptive beamforming for uplink. There are a couple of reasons for this. First, adaptive beamforming is traditionally used for reception in remote sensing, radar, and sonar systems. The adaptive beamforming techniques and algorithms that have been developed for reception readily lend themselves to wireless communications systems. Second, the spatial channel information is available on the uplink, thereby making it much easier to develop beamforming techniques and algorithms to deal with multipath and cochannel interference.

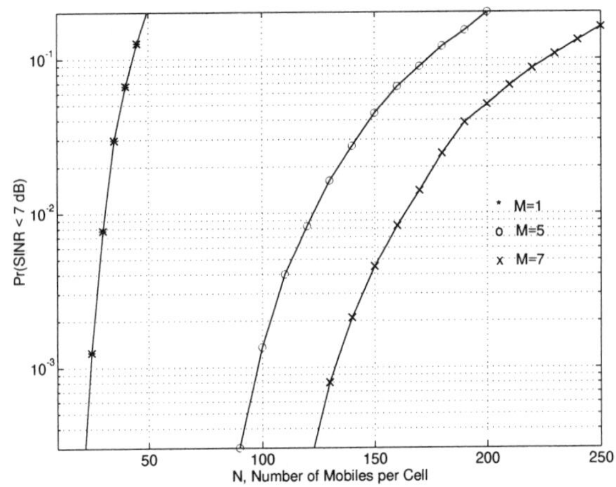

Figure 8.7 Downlink outage probability as a function of the number of antenna elements and cell loading . (From [17] ©1994 IEEE.)

8.4.1 Adaptive Criteria

In Chapter 3, we showed that the optimum criteria for beamforming are closely related to each other. That is, the optimum weights using different criteria are all given by the Wiener solution. These optimum criteria are as valid in communications as in other applications. In particular, these criteria are directly applicable if the objective of using an adaptive antenna is to increase the capacity of a mobile communications system. Although different forms of criteria may appear in the open literature, it can be shown that they all stem from the Wiener-Hopf equation. This is because the Wiener solution provides the upper limit on the theoretical adaptive beamforming steady-state performance.

8.4.2 Adaptive Algorithms

The common adaptive algorithms that have been investigated for beamforming in mobile communications include the LMS algorithm [8,9,19–21], the SMI technique [9,11,19,20,22–24] and RLS algorithm [20,21,25]. The principles of these algorithms were presented in Chapter 3. Here some implementation issues will be discussed.

The LMS algorithm is simple to implement. However, it has its drawbacks. In addition to its inherent weak characteristics, which were mentioned in Chapter 3 the dynamic range over which it operates is quite limited. In the example given by Winters [8], the permissible dynamic range is only 20 dB for a specific scenario. Since the received signals in a mobile radio system vary by more than 20 dB, power control is required if the LMS algorithm is to be used. Alternatively, the normalized LMS algorithm [26] can be used to overcome the dynamic range limitation.

The SMI approach offers a relatively fast convergence rate. In practice, the signal environment varies with time due to fading. Thus, the basestation must continually update the weight vector to adapt to the change in the channel. The updating rate depends on the fading rate of the channel. Although the SMI approach can be shown, in theory, to converge more rapidly than the LMS algorithm, the practical difficulties that are associated with the SMI algorithm approach should be appreciated. There are two major problems of its own: (1) increased computational complexity that cannot be easily overcome through the use of very-large-scale integration, and (2) numerical instability resulting from the use of finite-precision arithmetic and the requirement of inverting a large matrix.

An important feature of the RLS algorithm is that the inversion of the covariance matrix is replaced at each step by a simple scalar division. This feature reduces the computational complexity while maintaining a similar performance. The convergence rate of the RLS algorithm is typically an order of magnitude faster than that of the LMS algorithm, provided that the signal-to-noise ratio is high. The forgotten factor in the formulation of the RLS algorithm presented in Chapter 3 is very dependent on the fading rate of the channel. For a stationary channel, the forgotten factor can be set to be unity. For a fast-fading channel, a value slightly below unity should be chosen. A value of 0.95 was reported to be reasonable [20].

In addition to the above three adaptive algorithms, other adaptive techniques have also been applied to beamforming in mobile communications. These techniques include the conjugate gradient method [27], the eigenanlysis algorithm [23], the method based on rotational invariance [25], the linear least squares error (LLSE) algorithm [11], and the Hopfield neural network [28, 29]. All these methods have been proposed to either overcome the shortcomings or improve the performance of the LMS, SMI, or RLS algorithms when used for beamforming in mobile communications.

8.4.3 Reference Signals

If an explicit reference signal is available in a communications system, it should be used as much as possible for less complexity, high accuracy, and fast convergence. Explicit reference can be divided into two categories: spatial reference and temporal reference. Spatial reference is mainly referred to as the AOA information of the desired signal and its multipath components. A temporal reference signal may be a pilot signal that is correlated with the wanted signal, a special sequence in a packet carried by the wanted signal, or a known PN code in a CDMA system.

Spatial Reference

A number of approaches based on AOA information have been proposed for canceling cochannel interference [11, 25, 30, 31]. AOA information is usually obtained by applying a particular AOA estimation technique to the received array signals. A variety of AOA estimation techniques are available, which include wavenumber estimation techniques and parametric estimation techniques. Wavenumber estimation techniques are solely based on the decomposition of a covariance matrix whose terms consist of estimates of the correlation between the signals at the elements of an array antenna. Some examples are the multiple signal classification (MUSIC) algorithm [32], the modified forward-backward linear prediction (FBLP) algorithm [33], the Principal Eigenvector Gram-Schmidt (PEGS) algorithm [34], and the approach of Estimation of Signal Parameters by Rotational Invariance Techniques (ESPRIT) [35]. However, all of these algorithms are not effective for coherent signals. The parametric estimation approach mainly comprises a variety of maximum likelihood estimation (MLE) techniques. In the MLE approach, a particular likelihood function is formulated for the given radio signals. The MLE estimates of desired parameters, such as angles of arrival, are the ones for which the likelihood function is maximized. These techniques require a high degree of computational complexity. An obvious drawback with AOA-based approaches is the requirement for array calibration, in addition to the extra processing load required for estimating the AOA.

Temporal Reference

IS-54 is a digital TDMA standard. In IS-54, there are three user slots in a time frame, each of which contains 324 bits, including a 28-bit synchronization sequence, 12-bit user identification sequence, and 260 data bits. In [36], it is proposed that the known 28-bit synchronization sequence is used to generate the reference signal. Although the synchronization sequence for a given time slot is the same for all users, it is different for each of the three time slots in a given frame. Because basestations operate synchronously, signals from other cells have a high probability of having different timing and being uncorrelated with the reference signal for the desired signal. Furthermore, the 12-bit user identification code can be used to verify that the correct user's signal has been captured. In this way, the basestation in a given cell is able to distinguish the desired user's signal from users in other cells. There is only one cochannel user in a cell. The capacity of the system can be improved by reducing the frequency-reuse factor and by increasing the SINR using adaptive beamforming.

In a GSM system, there is a training sequence in each data burst. This training sequence can be used to create a reference signal. If the discrete time channel between the transmitted symbols and received samples is independent of the sampling timing, one can construct an exact reference signal using a fixed filter [37]. If the channel varies with the sampling timing, the sampling instant is not synchronized with the symbol interval. In this case, the modulation and sampling can be approximated by a set of discrete time channels, corresponding to different sampling instants within the symbol interval. Therefore, an adaptive filter can be used to generate the reference signal [24].

A reference signal scheme has been proposed for a digital European cordless telephone (DECT) system [22]. In this scheme, a training sequence is implemented in each DECT data burst. Each mobile unit uses a dedicated training sequence so that the basestation can recognize the users by their training sequences. These training sequences can be used as reference signals for beamforming. Gold sequences that are 31 bits long are used as the training sequences, which are implemented in the B-field of the DECT frame. The B-field is the second part of the data field of the DECT frame. This scheme leads, of course, to fewer data bits available for speech data. As a result, the speech coding rate must be lowered.

3.4.4 Adaptive Beamforming for CDMA

For the past few years, adaptive beamforming for CDMA wireless communications has been drawing more and more attention from both the communications and antenna communities. There are a couple of reasons. First, CDMA standards for wireless communications systems are gaining acceptance from the industry, especially in North America. Second, as discussed in Chapter 3, adaptive beamforming is highly suitable for a CDMA system, because the spreading codes can be used as references for beamforming.

The approach presented in Section 3.5 has been used in [21, 25, 28, 29]. Results presented in these works show that a significant increase in performance and capacity is obtained by using adaptive antennas. For example, without considering the multipath effects, 100% increase in capacity at BER $= 10^{-2}$ has been achieved using a four-element adaptive antenna in a CDMA system with 129 Gold sequences of length 127 and a 9600 bps transmission rate [25]. In another study [21], a more realistic channel model was considered, in which both multipath and cochannel interference are included. It was claimed that 15- to 35- dB gain in SIR, depending on the number of users, was obtained using a eight-element adaptive antenna in a

CDMA system with a 255-chip PN sequence.

In the cases where the communications channel has a long profile of delay spread a joint spatiotemporal process may be effectively used to deal with both multipath fading and cochannel interference. Of course, the improvement in performance i at the expense of high system complexity. A simple and obvious way to carr out spatiotemporal processing is to cascade an adaptive beamformer with a rak receiver [38]. In the configuration given in [38], a bank of correlators is used fo code-filtering the array signals before carrying out adaptive beamforming, which i similar to the structure shown in Figure 6.11. The code filters are used to isolat cochannel user signals. In the approach given in [39], the adaptive beamformer i cascaded with a bank of digital filters. The digital filters are used both to model th channels and to cancel cochannel interference. A similar multistage configuration i given in [40]. Joint optimal processing has been proposed by several authors [41,42] in which two-dimensional adaptive filter is used to process the code-filtered signal to maximize the SINR at the output of the two-dimensional adaptive filter. In al these cases, considerable improvement in performance has been reported.

8.4.5 Blind Adaptive Beamforming

In the case where an explicit reference signal is not available, blind adaptive beam forming has to be used. The principles of a number of commonly used blind beam forming techniques have been presented in Chapter 5. In what follows, we wil summarize some typical blind beamforming results and performance for wireles communications.

Constant Modulus Algorithm

The CMA has been investigated for both compensating fading and canceling cochan nel interference. It has been applied to advanced mobile phone service (AMPS) an IS-54 signals [43], GMSK signals [44,45], and 16-QAM signals [46].

The implementation of the CMA in an AMPS system is straightforward, sinc AMPS is the existing analog FM standard adopted in North America. When th SIR is positive in decibels so that the algorithm is able to capture the desired signal significant improvement in terms of MSE has been reported, especially at low level of SIR [43].

Although IS-54 signals are not strictly constant modulus signals, at the cente of the pulse the modulus is constant due to pulse shaping. In IS-54, there are thre users operating in a time frame, and therefore a beam has to be switched over th three time slots. Since the array response to each user is different, an appropriat technique is required to carry out time-multiplexing for tracking the weights fo lifferent users. The results reported in [43] show improvement in BER when the CMA is used in a white Gaussian channel with carrier-to-noise ratio between 0 and 8 dB. Considerable improvement in BER was also obtained in a cochannel environment with SIR between 0 and 10 dB.

Experiments have been carried out by applying CMA to GMSK signals to evaluate the performance in a frequency-selective fading environment [45]. Figure 8.8 gives the BER performance of the CMA measured on two different courses. These results show remarkable improvement in BER when using adaptive antennas.

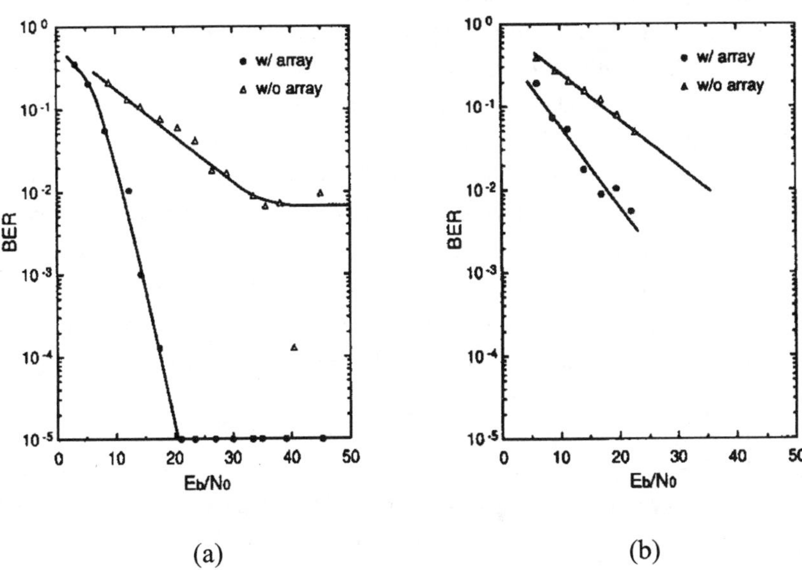

Figure 8.8 Comparison of experimental BER performance of a CMA adaptive antenna and an omnidirectional antenna. (From [45] ©1993 IEEE.)

Decision-Directed Algorithm

The decision-directed beamforming for wireless communications has been studied by a number of authors [47–49]. This approach can be implemented at low cost, since it is not computationally intensive and no array calibration is required. In a study given in [48], it is reported that the approach exhibits fast convergence, typically within 50 symbols. The algorithm locks on the desired signal with a probability of 99.9% at SIR levels as low as 1 dB. Cochannel rejection is typically more than 20 dB. The implementation was based on incoherent differential binary phase-shift keying (DBPSK) demodulation and the LMS algorithm. In another study given in [49], performance comparison was made between the decision-directed algorithm and the CMA, as well as the SCORE algorithm. Results show that the decision-directed algorithm converges much faster than other algorithms, achieving the theoretical performance limit after 50 symbol decisions.

Cyclostationary Algorithms

A cyclostationary algorithm has been developed for and applied to AMPS signals [50, 51]. AMPS signals exhibit the property of cyclostationarity, which arises due to the presence of the supervisory audio tone in an AMPS signal. The supervisory audio tone is used to avoid a false supervisory indication caused by cochannel interference. Three frequencies are allocated for supervisory audio tone signaling: 5970, 6000, and 6030 Hz. The basestation sends out a supervisory audio tone and the mobile sends back a supervisory audio tone. It the two are the same, the loop will be connected. If not, the basestation decides that the signal is interference. In [43], a spectral correlation technique has been developed to exploit this particular cyclostationary property of AMPS signals. The results reported in [43] show considerable improvement in MSE compared to the case of an omnidirectional antenna. The technique gives one-order improvement in MSE, even when the input SIR is negative in decibels. That is, there is no capturing difficulty, which is encountered in the application of the CMA.

We pointed out in Chapter 5 that the GMSK signals have cyclic frequencies at $\{\alpha_c = 2\Delta f \pm k\frac{B_r}{2}$, for $k = 0, 1, \cdots\}$. Therefore, cyclic beamforming can be applied to systems that use GMSK, such as GSM and DECT. The implementation requirements are that cochannel users must have slightly different carrier frequencies (e.g., 1-kHz separation). In [52], a study was carried out to investigate the use of cyclic adaptive beamforming (CAB) in a GSM system, as well as in a DECT system. Figures 8.9 and 8.10 give the SINR performance for a GSM system and a DECT system, respectively. In the GSM case, there are nine cochannel user signals, and eight of them are considered cochannel interference. Each cochannel signal has the same SNR of 10 dB and the same baud rate of 270.8 kbps. A 13-element linear array is used in a 120° sector cell. In the DECT case, the setting is similar, except that a 13-element circular array is used to cover the entire 360° angular range. In both cases, the capacity is increased by almost ninefolds. This number may be too optimistic, since the analysis did not include multipath effects.

8.5 ADAPTIVE BEAMFORMING FOR DOWNLINK

Up to this point, our discussion on the use of adaptive antennas in mobile communications has been centered on the uplink or reverse link (i.e., mobile to basestation)

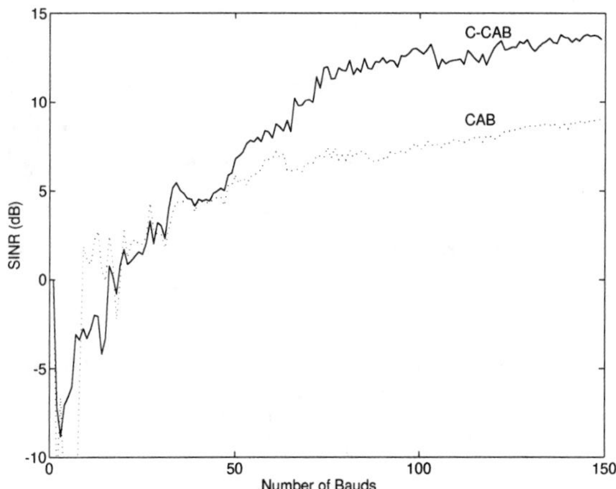

Figure 8.9 SINR performance of the CAB and constrained CAB techniques applied to a GSM system. (From [52] ©1994 Ho.)

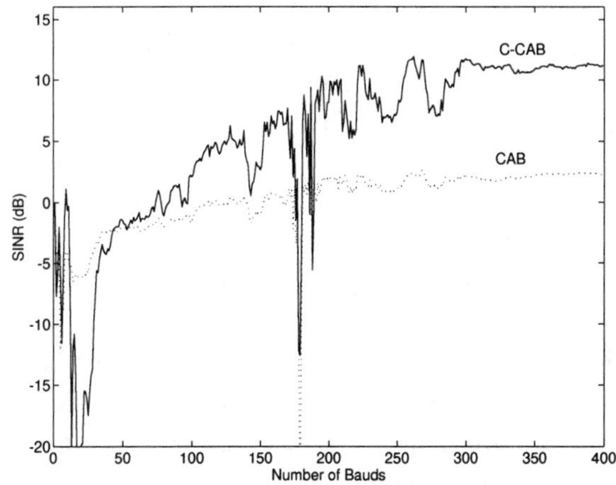

Figure 8.10 SINR performance of the CAB and constrained CAB techniques applied to a DECT system. (From [52] ©1994 Ho.)

case. In fact, it can be said that most of the research effort has been concentrated on the uplink case. Only in the last few years, has the difficulty associated with downlink beamforming been appreciated. This is why less research effort has been drawn to the downlink case.

As we mentioned earlier, the objective of adaptive beamforming for uplink is to maximize the SINR of the received desired signal. In the downlink or forward link (basestation to mobiles) case, the objective remains the same, although the means to achieve it may be different. That is, using the adaptive antennas, one has both to maximize the received signal strength at the desired mobile and to minimize the interference to other mobiles and adjacent basestations, thereby maximizing the downlink SINR.

If the transfer function of the channel at the downlink frequency is known, the downlink SINR can be maximized by multiplying the desired signal with a set of downlink weights. It can be shown that the weights are a scale version of the uplink weights, provided that the frequency for both the uplink and downlink is the same and that the channel is relatively static during the time separation between reception and transmission. Therefore, this concept of weight reuse can be applied to systems that employ the time-division duplexing (TDD) technique to time-share a single-frequency channel for transmission and reception in both directions [8]. For example, TDD is used in the following standards: CT2/CT2+ (Cordless Telephone 2), DECT, PHP (Personal Handy Phone), and DCS 1800.

In other standards such as IS-54, IS-95, and GSM, the frequency-division duplexing (FDD) technique is used. The concept of weight reuse is not applicable to these standards, because the separation between the downlink and uplink frequencies is so large that the corresponding channel transfer function is different to a large extent. The difference is basically due to the fact that the relative phases of the signals arrive via multiple paths and vary with changes in frequency. In the analysis given in [53], the results show that the downlink performance even degrades in a multipath environment when the uplink weights are used as the downlink weights for an FDD system. Therefore, techniques must be developed to derive the downlink weights.

The essence of the problem in downlink beamforming is how to estimate the channel transfer function. A feedback technique has been suggested by Gerlach and Paulraj [54]. In this approach, the basestation transmits probing signals at the downlink frequency. During the probing, normal information transmission is temporally halted. Each mobile in the cell measures its own responses to the probing signals and reports them back to the basestation. Based on the reported responses, the downlink channel transfer function is estimated for deriving the downlink weights. The concept in this approach is simple. However, it requires a complete redesign f protocols and signaling. Furthermore, due to the time required for probing, the pproach is only applicable in environments that change slowly.

Instead of using feedback, a mobile in the cell is required to directly transmit narrowband testing signal at the downlink frequency so that the basestation is ble to directly estimate the downlink channel transfer function and derive the ownlink weights. With this approach, it is not necessary to interrupt the normal plink transmission. However, it still requires a complete redesign of protocols and gnaling. Furthermore, additional hardware is required in the mobile unit.

A number of approaches based on AOA information have been proposed [31, 5, 56]. Their premise is that although the transfer function is different at uplink nd downlink frequencies, the AOAs remain relatively unchanged. Based on the OA information, the downlink weights are derived by maximizing the SINR. If he channel is a flat fading channel, the probability that a deep fade occurs at the esired mobile will be small, even though the transmission of the desired signal may e via different paths.

An interesting but simple technique is to use fixed multiple beams for both eception and transmission at the basestation [57]. The strongest beam in uplink will e used for downlink. The concept can be extended to the use of steerable multiple eams. On the uplink, the basestation determines the direction of the path on which he strongest component of the desired signal arrives at the basestation. On the ownlink, the basestation points a beam in the corresponding direction. Although his approach is not optimal, the SINR level at the mobile can be improved. Since a arrow beam is pointed, either directly or indirectly, at the mobile, the probability hat the mobile is in a deep fade is extremely small. Furthermore, at the basestation, ower is not a critical resource and a relatively high-power beam can be used to oost the SINR level if necessary. This is in contrast to the uplink case in which mobile transmits using an omnidirectional antenna with very limited power and herefore the desired signal components arrive via multiple paths at the basestation nd must be used to obtain a desirable SINR level.

References

[1] R. T. Compton, R. J. Huff, W. G. Swarner, and A. A. Ksienski, "Adaptive arrays for communication systems: An overview of research at the Ohio State University," *IEEE Trans. Antennas Propagat.*, vol. AP-24, pp. 599–607, 1976.

[2] R. T. Compton, "An adaptive antenna in a spread spectrum communication system," *Proc. IEEE*, vol. 66, pp. 289–398, March 1978.

[3] J. H. Winters, "Increased data rate for communication systems with adaptive antennas," in *Proc. IEEE Int. Conf. Comm.*, June 1982.

[4] Y. S. Yeh and D. O. Reudink, "Effecient spectrum utilization for mobile radio system using space diversity," in *IEE Int. Conf. Radio Spectrum Conservation Techniques*, London, pp. 12–16, 1980.

[5] Y. S. Yeh and D. O. Reudink, "Effecient spectrum utilization for mobile radio systems using space diversity," *IEEE Trans. Comm.*, vol. COM-30, pp. 447–455, March 1982.

[6] V. M. Bogachev and I. G. Kiselev, "Optimum combining of signals in space-diversity reception," *Telecomm. Radio Eng.*, vol. 34/35, pp. 83–85, Oct. 1980.

[7] M. J. Marcus and S. Das, "The potential use of adaptive antennas to increase land mobile frequency reuse," in *IEE Int. Conf. Radio Spectrum Conservation Techniques*, University of Birmingham, U.K., pp. 113–117, 1983.

[8] J. H. Winters, "Optimum combining in digital mobile radio with co-channel interference," *IEEE Trans. Veh. Technol.*, vol. 33, pp. 144–155, Aug. 1984.

[9] R. G. Vaughan, "On optimum combining at the mobile," *IEEE Trans. Veh. Technol.*, vol. 33, pp. 181–188, Nov. 1984.

[10] S. C. Swales, M. A. Beach, D. J. Edwards, and J. P. McGeehan, "The performance enhancement of multibeam adaptive basestation antennas for cellular land mobile radio systems," *IEEE Trans. Veh. Technol.*, vol. 39, pp. 56–57, Feb. 1990.

[11] S. Anderson, M. Millnert, M. Viberg, and B. Wahlberg, "An adaptive array for mobile communications systems," *IEEE Trans. Veh. Technol.*, vol. 40, pp. 230–236, Feb. 1991.

[12] M. Barrett and R. Arnott, "Adaptive antennas for mobile communications," *Electron. Comm. Eng. J.*, vol. 6, pp. 203–214, Aug. 1994.

[13] R. C. French, "The effects of fading and shadowing on channel reuse in mobile radio," *IEEE Trans. Veh. Technol.*, vol. VT-28, pp. 171–181, Aug. 1979.

[14] R. Muammar and S. Gupta, "Co-channel interference in high capacity mobile radio system," *IEEE Trans. Comm.*, vol. COM-30, pp. 1973–1978, Aug. 1982.

[15] M. B. Pursley, "Performance evaluation for phase-coded spread spectrum multiple-access communications with random signature sequences," *IEEE Trans. Comm.*, vol. COM-25, Aug. 1977.

[16] J. C. Liberti and T. S. Rappaport, "Analytical results for capacity improvements in CDMA," *IEEE Trans. Veh. Technol.*, vol. 43, pp. 680–690, Aug. 1994.

[17] A. F. Naguib, A. Paulraj, and T. Kailath, "Capacity improvement with base-station antenna arrays in cellular CDMA," *IEEE Trans. Veh. Technol.*, vol. 43, pp. 691–698,

Aug. 1994.

[18] K. S. Gilhousen et al., "On the capacity of a cellular CDMA system," *IEEE Trans. Veh. Technol.*, vol. 40, pp. 303–312, May 1991.

[19] J. H. Winters, "Signal acquisition and tracking with adaptive arrays in the digital mobile radio system IS-54 with flat fading," *IEEE Trans. Veh. Technol.*, vol. 42, pp. 377–384, Nov. 1993.

[20] J. Fernandez, I. R. Corden, and M. Barrett, "Adaptive array algorithms for optimal combining in digital mobile communications systems," in *IEE Eighth Int. Conf. Antennas and Propagat.*, Heriot-Watt University, U.K., 1993.

[21] G. Tsoulos, M. A. Beach and S. C. Swales, "Adaptive antennas for third generation DS-CDMA cellular systems," in *Proc. IEEE VTC'95*, Chicago, pp. 45–49, Aug. 1995.

[22] J. Wigard, P. E. Mogensen, F. Frederiksen, and O. Norklit, "Evaluation of optimum diversity combining in DECT," in *Sixth IEEE Int. Symposium on Personal, Indoor and Mobile Radio Communications*, Toronto, pp. 507–511, Sept. 1995.

[23] X. Wu, A. M. Haimovich, "Adaptive arrays for increased performance in mobile communications," in *Sixth IEEE Int. Symposium on Personal, Indoor and Mobile Radio Communications*, Toronto, pp. 653–657, Sept. 1995.

[24] E. Lindskog, "Making SMI-beamforming insensitive to the sampling timing for GSM signals," in *Sixth IEEE Int. Symposium on Personal, Indoor and Mobile Radio Communications*, Toronto, pp. 664–668, Sept. 1995.

[25] Y. Wang and J. R. Cruz, "Adaptive antenna arrays for cellular CDMA communication systems," in *Proc. IEEE Int'l. Conf. Acoustics, Speech and Signal Processing*, Detroit, pp. 1725–1728, 1995.

[26] S. Haykin, *Adaptive Filter Theory*, Prentice Hall, Englewood Cliffs, NJ, 1991.

[27] S. Choi and T. Sarkar, "Adaptive antenna array utilizing the conjugate gradient method for multipath mobile communications," *Signal Processing*, vol. 29, pp. 319–333, 1992.

[28] B. Quach, H. Leung, T. Lo, and J. Litva, "Hopfield network approach to beamforming in spread spectrum communications," in *IEEE Proc. Seventh SP Workshop on Statistical Signal & Array Processing*, pp. 409–412, June 1994.

[29] A. Sandhu, T. Lo, H. Leung, and J. Litva, "A Hopfield neurobeamformer for spread spectrum communications," in *Sixth IEEE Int. Symposium on Personal, Indoor and Mobile Radio Communications*, Toronto, Sept. 1995.

[30] B. Ottersten, R. Roy, and T. Kailath, "Signal waveform estimation in sensor array processing," in *23rd Asilomar Conf. Signals, Systems and Computers*, 1989.

[31] P. Zetterberg and B. Ottersten, "Spectrum efficiency of base-station antenna array system for spatially selective transmission," in *Proc. IEEE VTC'94*, pp. 1517–1521, 1994.

[32] R. Schmidt, "Multiple emitter location and signal parameter estimation," in *Proc. of the RADC Spectrum Estimation Workshop, RADC-TR-79-63*, Rome Air Development Center,Rome, NY, 1979.

[33] D. W. Tufts and R. Kumaresan, "Estimation of frequencies of multiple sinusoids: Making linear prediction perform like maximum likelihood," *Proc. IEEE*, vol. 70, pp. 975–989, Sept. 1982.

[34] W. F. Gabriel, "A high-resolution target-tracking concept using spectral estimation techniques," Technical Report 8797, NRL, May 1984.

[35] A. Paulraj, R. Roy, and T. Kailath, "Estimation of signal parameters by rotational invariance techniques ESPRIT," in *19 Asilomar Conf. Signals, Systems and Computers*, 1985.

[36] J. H. Winters, J. Salz, and R. D. Gitlin, "The impact of antenna diversity on the capacity of wireless communication systems," *IEEE Trans. Comm.*, vol. 42, pp. 1740–1750, Feb./March/Apr. 1994.

[37] Y. Ogawa, Y. Nagashima, and K. Itoh, "An adaptive antenna system for high-speed digital mobile communications," *IEICE Trans. Comm.*, vol. E75-B, pp. 413–421, May 1992.

[38] A. F. Naguib and A. Paulraj, "Performance of CDMA cellular networks with base-station antenna arrays," in C. G. Gunther, ed., *Mobile communications—Advanced Systems & Components*, Springer-Verlag, pp. 87–100, March 1994.

[39] R. Kohno, H. Imai, M. Hatori, and S. Pasupathy, "Combination of an adaptive array antenna and a canceller of interference for direct-sequence spread-spectrum multiple-access system," *IEEE . Select. Areas Comm.*, vol. 8, pp. 675–682, May 1990.

[40] V. Ghazi-Moghadam and M. Keveh, "Interference cancellation using adaptive antenns," in *Sixth IEEE Int. Symposium on Personal, Indoor and Mobile Radio Communications*, Toronto, pp. 936–939, Sept. 1995.

[41] H. Iwai, T. Shiokawa, and Y. Karasawa, "An investigation of space-path hybrid diversity scheme for base station reception in CDMAmobile radio," *IEEE J. Sel. Areas Comm.*, vol. SAC-12, pp. 962–969, June 1994.

[42] R. Kohno, N. Ishii, and M. NagatsU.K.a, "A spatially and temporally optimal multi-user receiver using an array antenna for DS/CDMA," in *Sixth IEEE Int. Symposium on Personal, Indoor and Mobile Radio Communications*, Toronto, pp. 950–954, Sept. 1995.

[43] P. Petrus and J. H. Reed, "Least squares CM adaptive arrays for cochannel interference rejection for AMPS and IS-54 signals," in *Wireless'95*, Calgary, July 1995.

[44] T. Ohgane et al., "An implementation of a CMA adaptive array for high speed GMSK transmission in mobile communications," *IEEE Trans. Veh. Technol.*, vol. 42, pp. 282–288, Aug. 1993.

[45] T. Ohgane et al., "BER performance of CMA adaptive array for high speed GMSK mobile communication—A description of measurements in Central Tokyo," *IEEE Trans. Veh. Technol.*, vol. 42, pp. 484–490, Nov. 1993.

[46] N. Kikuma et al., "Consideration on performance of the CMA adaptive array antenna for 16-QAM signals," in *Sixth IEEE Int. Symposium on Personal, Indoor and Mobile*

Radio Communications, Toronto, pp. 677–681, Sept. 1995.

[47] J. Yang, S Daas, and A. Swindlehurst, "Improved signal copy with partial known or unknown array response," in *Proc. IEEE Int'l. Conf. on Acoustics, Speech and Signal Processing*, Adelaide, Australia, pp. IV–265–IV268, 1994.

[48] V. Kezys and J. Litva, "A versatile intelligent antenna testbed for wireless applications," in *Wireless'95 Proc.*, Calgary, Alberta, 1995.

[49] A. Swindlehurst, S. Daas, and J. Yang, "Analysis of a decision directed beamformer," *IEEE Trans. Signal Processing*, vol. 43, pp. 2920–2927, Dec. 1995.

[50] R. He and J. Reed, "Spectral correlation of AMPS signals and its application to interference rejection," in *Proc. MILCOM'94*, 1994.

[51] P. Petrus and J. H. Reed, "Cochannel interference rejection for MPSsignals using spectral correlation properties and an adaptive array," in *Proc. IEEE VTC'95*, Chicago, pp. 30–34, Aug. 1995.

[52] R. Ho, "Implementation of cyclic beamforming techniques on mobile communication systems," Master's thesis, McMaster University, Hamilton, Ont., 1994.

[53] T. Ohgane, "Spectral efficiency improvement by base station antenna pattern control for land mobile cellular systems," *IEICE Trans. Comm.*, vol. E77-B, pp. 598–605, May 1995.

[54] D. Gerlach and A. Paulraj, "Adaptive transmitting antenna arrays with feedback," *IEEE Signal Processing Letters*, vol. 1, pp. 150–152, Oct. 1994.

[55] G. Xu and H. Liu, "An effective transmission beamforming scheme for frequency-division-duplex digital wireless communication systems," in *Proc. IEEE Int'l. Conf. on Acoustics, Speech and Signal Processing*, Detroit, pp. 1729–1732, 1995.

[56] C. Farsakh and J. A. Nossek, "Cahnnel allocation and downlink beamforming in an SDMA mobile radio system," in *Sixth IEEE Int. Symposium on Personal, Indoor and Mobile Radio Communications*, Toronto, pp. 687–691, Sept. 1995.

[57] R. Bernhardt, "The use of multiple-beam directional antenna in wireless messaging systems," in *Proc. IEEE VTC'95*, Chicago, IL, pp. 858–861, Aug. 1995.

Chapter 9

Adaptive Beamforming in Indoor and Data Communications

In the preceding chapter, the discussion was centered on adaptive beamforming in land mobile communications. Although adaptive beamforming in both indoor and data communications is similar to that in mobile communications in many aspects, there are also differences. Therefore, a separate chapter is needed to describe adaptive beamforming in both indoor and data communications. In the first section of this chapter, we will discuss the use of adaptive antennas in indoor communications. In the following section, we will address adaptive beamforming in mobile data communications.

9.1 INDOOR COMMUNICATIONS

9.1.1 Characteristics of Indoor Communications

Although indoor radio propagation is dominated by the same mechanisms as outdoor (reflection, diffraction, and scattering), an indoor communications channel differs from a mobile channel in a number of aspects:

1. The characteristics of an indoor channel are greatly dependent on specific features, such as the layout of the building, the construction materials, and the building type. For example, the number of multipath components in each impulse response profile is a random variable and its mean value depends on the types of building.
2. Path loss in an indoor environment is very severe and also very dynamic, changing appreciably over a very short distance. Simple path loss rules are successful in predicting a mobile channel, but not an indoor channel.
3. Generally, an indoor channel is nonstationary in time. Temporal variations are due to the motion of people and equipment around antennas of both basestations and portables. For example, the signal levels changes greatly

when a portable unit moves from a room to a hallway, or when the door of the room in which the portable unit is located is being open or closed.

4. The distances covered are relatively small. That is, the cell size for an indoor communications network is much smaller than that for mobile communications. The relatively small propagation distance makes it difficult to ensure far-field radiation for the entire coverage region. The variability of the environment is much greater for a small transmission-reception range. Furniture such as bookcases, file cabinets, and desks may become large obstacles that introduce shadowing effects.
5. In a mobile environment, fading is relatively narrowband, with coherence bandwidths on the order of 100 kHz. In contrast, fading occurring in an indoor channel may have a bandwidth as wide as 10 MHz. In this case, a rake receiver may be completely ineffective.
6. The length of the impulse response, or delay spread for an indoor channel is usually on the order of tens of nanoseconds, which is relatively small compared to that for a mobile channel (on the order of tens of microseconds).

A comprehensive discussion on indoor channels can be found in [1]. In general, the propagation conditions in an indoor environment are more severe than those in a mobile environment.

Traditionally, it is adequate to characterize an indoor channel in terms of the power level and temporal characteristics. However, the spatial characteristics are equally, if not more, important when beamforming is concerned. To quantify the spatial characteristics, we can define a number of quantities that are similar to those for multipath delay. The power-weighted angular profile as

$$f(\phi) = \sum_i \beta_i^2 \delta(\phi - \phi_i) \tag{9.1}$$

where β_i^2 denotes the power of the ith multipath component which arrives at the basestation at an angle of ϕ_i. The power-weighted average AOA is given by

$$\overline{\phi} = \frac{\sum_i \beta_i^2 \phi_i}{\sum_i \beta_i^2} \tag{9.2}$$

and the rms angular spread is given by

$$\sigma_\phi = \sqrt{\frac{\sum_i \beta_i^2 \left(\phi_i - \overline{\phi}\right)}{\sum_i \beta_i^2}} \tag{9.3}$$

In a mobile channel, the angular spread is relatively small. Multipath is caused mainly by the local scatterers which are the large objects such as houses and buildings surrounding the mobile unit [2]. Because the ratio of the distance between a basestation and a mobile (several kilometers) to the radius of scatterers (about 100 wavelengths), the angular sector within which multipath signals arrive is small. As a result, the angular spread is also small. In contrast, the angular spread in an indoor channel is relatively large. In an indoor environment, the local scatterers are objects within a building, such as walls, partitions, doors, desks, and shelves. The distance between a basestation and a portable unit is small. It is very possible that a basestation is within the radius of scatterers of a particular portable unit. Thus, the ratio of the distance between the basestation and a portable to the radius of scatterers is relatively large, thereby resulting in a large angular spread. For example, Figure 9.1 shows two sets of indoor AOA measurements [3]. The antenna responses to the incoming signals at different time intervals, shown as beam patterns in linear scale, were constructed from the measured data using the process outlined in [3]. The patterns in the figure were formed over a time period of 250 ns with intervals of 25 ns. The largest lobe in each figure corresponded to the AOA of the direct path of the transmitted signal. Other large lobes in the patterns indicated the AOAs of the multipath components that arrived at the receiving antenna at different time intervals in a relatively short time frame. It can be observed that strong multipath components arrived at the receiver in different directions from 0° to 360°, indicating a large angular spread. Furthermore, even though the two measurements shown in the figure were carried out only 2.5 wavelengths apart, the AOAs of multipath components were quite different. That is, the received signal changes greatly even if the change in the distance between the basestation and the portable is small.

The most significant impact of a large angular spread is that it makes sectorization of cells ineffective and perhaps even counterproductive. Therefore, omnidirectional (on the horizontal plane) antennas should be used in an indoor wireless communications network. If adaptive beamforming is to be used, an antenna array that is able to transmit and receive in any direction between 0° and 360° is required. A circular array is a very good candidate in that it is able to form a beam in any direction and the shape of the beam does not depend on the beam pointing direction.

9.1.2 Capacity with Adaptive Beamforming

Just as in the mobile wireless communications network, adaptive beamforming can also be used to increase the capacity of an indoor wireless communications network by suppressing cochannel interference so that multiple users can use the same channel simultaneously within a cell. However, since the propagation environments are different, the performance of an adaptive antenna in indoor communications will be

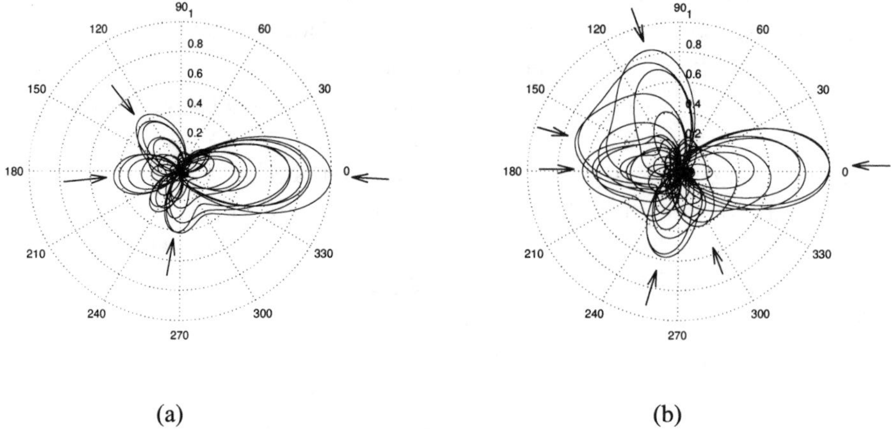

Figure 9.1 Antenna responses to the incoming signals at different time intervals; (a) and (b) are the results derived from two measurements were made only one foot apart.

different than that in mobile communications. Therefore, Winters has carried out an analysis to determine the capacity performance of an indoor system [4]. In what follows, a summary of his study is presented. The following scenario is assumed:

1. Multipath fading is independent Rayleigh.
2. The basestation antenna array consists of M elements.
3. There are K simultaneous users per channel, all with a BER less than 10^{-3}.
4. The SNR per antenna element for the desired signal is Γ_d.
5. The INR per antenna element for the ith cochannel interferer is Γ_i.
6. There are N frequency channels within a cell.

To determine the capacity of a multiple-users-per-channel system, the following approach is taken:

1. Determining the probability, P_K, that a given number of simultaneous users can use the same frequency channel for a given level of SNR and M antenna elements;
2. Determining the probability, $P_{K+1|K}$, that with K simultaneous users, another user can be added to the channel;
3. Determining the capacity of a system with a 1% blocking probability based on $P_{K+1|K}$.

Since it is extremely difficult to analytically determine P_K when $K > 1$, P_K has to be evaluated based on Monte Carlo simulation. Based on a large number of simulated cases, P_K is calculated as the proportion of cases in which all cochannel users have a BER less than 10^{-3}. In [4], it has been observed that the required level of SNR Γ_d must be increased as the number of cochannel users, K, increases, and that for $K = M$ cochannel users with high probability, it is required that Γ_d be increased by up to 20 dB. $P_{K+1|K}$ can be determined by noting that the BER for each of the existing K users can only be increased by adding an additional interferer. Thus, the cases where BER $< 10^{-3}$ with $K+1$ users are a subset of the cases where BER $< 10^{-3}$ with K users. $P_{K+1|K}$ is simply $\frac{P_{K+1}}{P_K}$. The call blocking probability for a single channel with capacity K is defined as the probability that the Kth user cannot be added to the system; that is, for a one-channel system $(N - 1)$,

$$p_B = 1 - P_{K+1|K} = \frac{P_{K+1}}{P_K} \tag{9.4}$$

The capacity is then the maximum number of simultaneous users for a particular value of p_B and M antenna elements. Figure 9.2 [4] shows the capacity as a function of Γ_d for $p_B = 1\%$. It can be observed that the required increase in Γ_d for each additional user becomes smaller as the number of antenna elements increases. For example, for five cochannel users with a six-element basestation antenna, it is required that $\Gamma_d = 17$ dB, whereas with a nine-element antenna, it is required that $\Gamma_d = 5$ dB. It should be noted that for M cochannel users, a substantial increase in Γ_d is necessary compared with the single-user case.

For $N > 1$, the system capacity will normally be increased by N-fold. However, if DCA is implemented, the capacity can indeed be increased more than N-fold. To evaluate the capacity, let us consider an N-channel system. We assume for simplicity that all channels must have K users before the $(K+1)^{\text{th}}$ additional user is added to the system. This is a worst-case assumption, since the capacity is actually greater if the number of users in each channel is nonuniformly distributed. If in $N - l$ channels there are K users per channel and in l channels there are $K+1$ users per channel, the total number of users is $K \times (N - l) + (K+1) \times l = NK + l$. The blocking probability for the $KN + l + 1$ user is given by [4]

$$p_B = \left(1 - P_{K+1|K}\right)^{N-l} \left(1 - P_{K+2|K+1}\right)^{l} \tag{9.5}$$

Figure 9.3 shows the system capacity as a function of Γ_d for an eight-channel system with $p_B = 1\%$. It can be observed that an M-fold increase in capacity is achievable with M-element basestation antennas if Γ_d is increased by as much as 20 dB. As the number of antenna elements increases, the required increase in Γ_d decreases. It should also be noted that as the number of channels increases for the same value of p_B, the required Γ_d decreases.

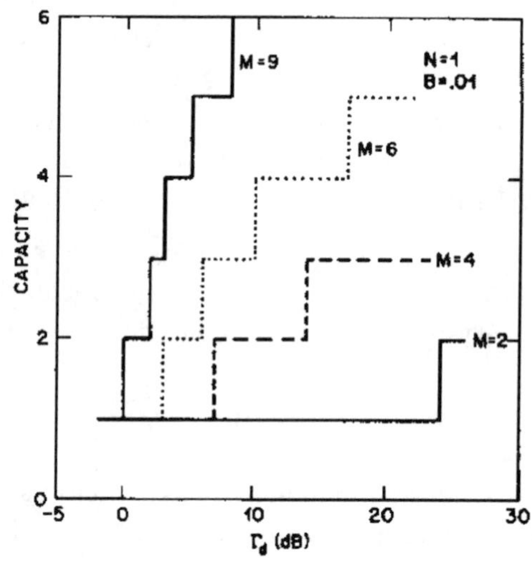

Figure 9.2 System capacity as a function of Γ_d for a single-channel system with an M-element basestation antenna and $p_B = 1\%$. (From [4] ©1987 IEEE.)

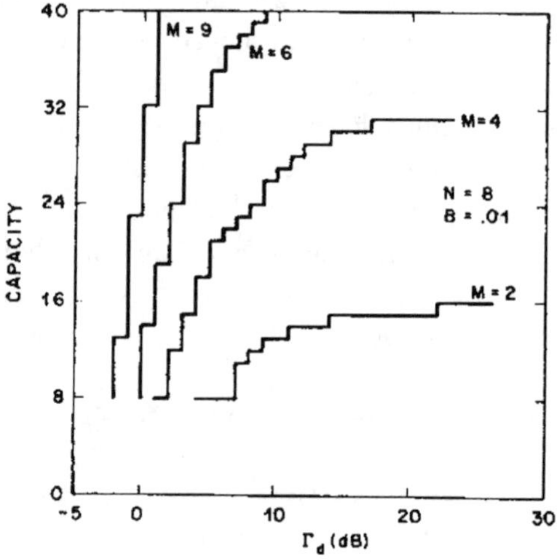

Figure 9.3 System capacity as a function of Γ_d for an eight-channel system with an M-element basestation antenna and $p_B = 1\%$. (From [4] ©1987 IEEE.)

9.1.3 Voice and Low-Data-Rate Indoor Communications

In a study given in [5], a linear adaptive antenna array with uniformly spaced elements is used in a TDMA indoor system. In the study, perfect synchronization and carrier recovery is assumed. Both the Rayleigh flat-fading channel model and the Racian channel model are considered. Other simulation parameters are listed in Table 9.1. The LMS algorithm is used to carry out adaptive beamforming. A training signal, which is a PN sequence, is used as the reference signal. The results reported in [5] show that there is considerable gain from using adaptive antennas. For example, using a two-element array over a Rayleigh flat-fading channel and in the presence of eight equal-power cochannel interferers at an average SIR = 5 dB, an order of improvement in BER (about 10^{-2}) has been observed compared to the case of a single antenna (about 10^{-1}). It has also been observed that the adaptive antenna performs better over a Racian channel than a Rayleigh channel. The results show that two-order improvement in BER has been achieved at a 5-dB ratio of the power in the line-of-sight component to that in the multipath components.

Table 9.1
Simulation Parameters Used in [5].

Carrier frequency	1.7 GHz
Doppler frequency	6–9 Hz
Average fade duration	3.8 – 7.2 ms at -20 dB
Data rate	800 kbps (400,000 symbols/second)
Sampling rate	one sample per symbol
Modulation	$\frac{\pi}{4}$-QPSK

In another study presented in [6], an adaptive circular array is used to combat multipath and cochannel interference in indoor communications. In the simulation analysis, multipath components are assumed to be discrete rays arriving in different distinct directions. Other simulation parameters are listed in Table 9.2. The LMS algorithm is also used in this particular study. A pilot training signal is first used to derive the beamforming weights. After the algorithm converges to a desired state (i.e., the array locks on the desired signal), the system is switched to the decision-directed mode to continue tracking the desired signal. Two cases of performance evaluation are given in [6]. In the first case, it is assumed that there are only multipath components in the desired signal. A number of observations have been obtained:

1. The symbol error rate (SER) decreases as the number of antenna elements increases.
2. The SER is lower for the case in which there is a dominant member among all the multipath components than for the case in which all the multipath components are equal in power. A dominant component normally has a relatively high level of SNR, resulting in a better SER performance.
3. With a large number of antenna elements, the SER performance is less sensitive to both the AOA of paths and the power difference between the multipath components.

In the second case, only cochannel interference is considered. In particular, there are a desired signal and a cochannel interference. The results show that even with two-elements, the SER can be improved considerably. For example, an SER of 10^{-2} is achieved using a two-element adaptive antenna at SNR = 8 dB, whereas the SER is 0.5 using a single antenna. In general, the SER performance is more sensitive to SIR for a small rather than large number of antenna elements. The SER performance deteriorates as the angular separation between the desired signal and the interferer decreases.

Table 9.2
Simulation Parameters Used in [6]

Carrier frequency	1.0 GHz
Doppler frequency	maximum 50 Hz
Data rate	200 kbps (100,000 symbols/second)
Sampling rate	one sample per symbol
Modulation	$\frac{\pi}{4}$-QPSK

The evaluation of adaptive beamforming based on the LMS algorithm in a indoor communications has been carried out under a condition of imperfect power control [7]. A number of findings that have been reported from this particular study are summarized as follows:

1. The LMS algorithm performs very well when the SIR is greater than -20 dB and log(BER) is nearly inversely proportional to the number of antenna elements.
2. The LMS algorithm is able to track the desired signal even for a low correlation between the reference signal and the desired signal.
3. The BER densities are well described by a log-gamma density function.

4. With a frequency-reuse factor of unity, adaptive beamforming is an effective means of increasing system capacity, and in this particular case, power control seems unnecessary.

It is interesting to examine the BER distribution. Under time-variant conditions, the instantaneous BER at the beamformer output fluctuates. Therefore, the BER distribution may provide more insight into the BER performance of the system. Figure 9.4 shows the BER distributions for different combinations of the numbers of cochannel users and the levels of interference-to-signal ratio (ISR, the inverse of SIR). The results have been obtained using a 16-element array at a -2-dB SNR. It can be observed that as the number of cochannel users increases or the level of ISR increases, the BER distribution shifts to the right. It appears that as the mean value of the BER decreases, the variance of the BER increases.

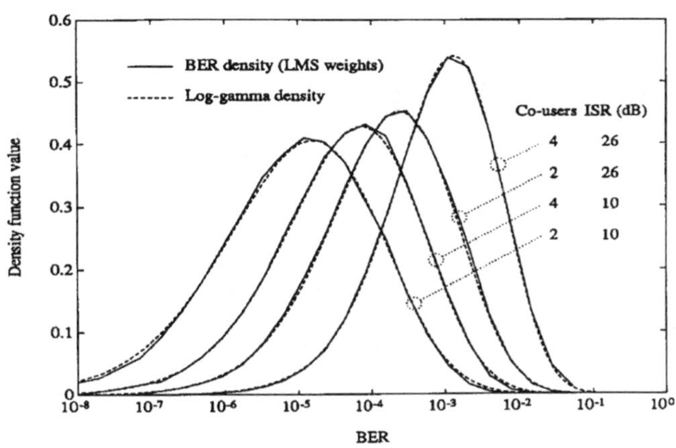

Figure 9.4 BER distribution for different combinations of the numbers of cochannel users and the ISR levels. (From [7] ©1994 Bree.)

In Figure 9.5(a), the BER performance is given as a function of the number of antenna elements. This plot can be used to determine the number of antenna elements required to achieve a certain level of BER for a given value of SNR. The results have been obtained under the conditions of a single user moving at a speed of 2 m/s over a Rayleigh flat-fading channel without cochannel interference. Figure 9.5(b) presentes, in a different way, the same results as those shown in Figure 9.5(a). The presentation is given in terms of the number of antennas as a function of SNR. For example, 10 elements are required to achieve a 10^{-4} BER at a 0-dB SNR.

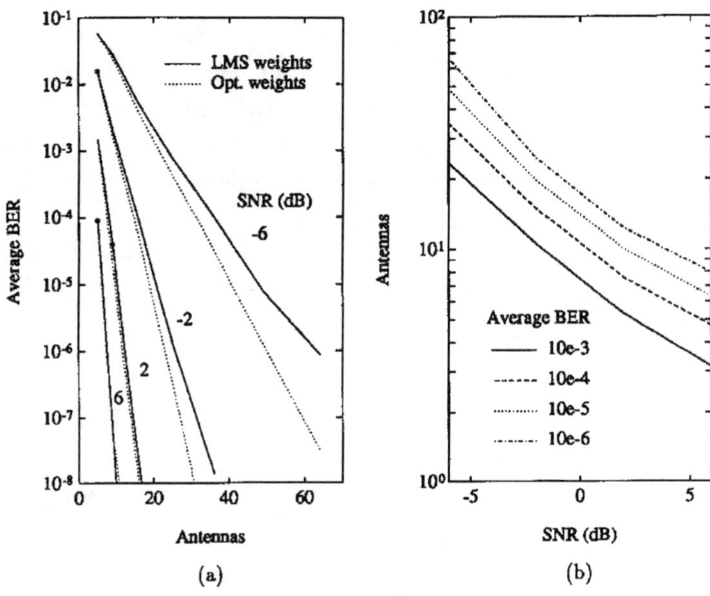

Figure 9.5 (a) Average BER as a function of the number of antenna elements for different levels of SNR; (b) number of antenna elements as a function of SNR for different values of BER. (From [7] ©1994 Bree.)

9.1.4 High-Data-Rate Indoor Communications

There are two major factors that limit the performance of a high-data-rate indoor communications system: the received signal power margin and delay spread. The received signal power margin is referred to as the difference, in decibels, between the SNR at the receiver and the SNR required to achieve a given BER.

The SNR at the receiver is a function of a number of parameters:

$$\mathrm{SNR} = P_t + G_t + G_r + 166 - 20\log f - L_p - T_n - \mathrm{NF} - 10\log B \quad (\mathrm{dB}) \tag{9.6}$$

where

P_t = transmission power
G_t = transmitting antenna gain
G_r = receiving antenna gain
f = frequency in GHz
L_p = propagation loss
T_n = system noise temperature
B = bandwidth
NF = receiver noise figure

Keeping other parameters fixed, a larger value of B results in a lower level of SNR. Since a given BER requires a certain level of SNR, B and hence the bit rate are limted. For example, if the following typical values are considered:

P_t = 16 dB
G_t = 3 dB
G_r = 3 dB
λ = 20 GHz
L_p = 60 dB
T_n = 25 dB at room temperature
NF = 6 dB

the level of SNR at the receiver is $(71 - 10\log B)$ dB. If we assume that a BER of 10^{-8} is achievable with coding, the required SNR is 8.7 dB for coherent detection of BPSK, resulting in a signal power margin of $(63.3 - 10\log)B$ dB. Therefore, the maximum data rate would be about 2.0 Mbps.

The delay spread in an indoor environment is on the order of 100 ns [1, 8, 9]. Without the use of an equalizer, a digital communications system requires the magnitude of rms delay spread to be less than 10% of the symbol period to achieve a BER of less than 10^{-8}. That is, the maximum data rate that can be achieved with a delay spread on the order of 100 ns is about 1 Mbps.

In order to achieve a data rate on the order of tens to hundreds of Mbps, one has to increase the SNR at the receiver, to reduce the delay spread, or to carry out both together. One may consider increasing the transmission power level. However, this may not necessarily be a practical solution. For example, achieving a 200-Mbps data rate requires a 20-dB increase in transmission power; that is, if 1W transmission power is required to achieve 2 Mbps, 100W are needed to achieve 200 Mbps! Equalization or multicarrier processing can be used to deal with the delay spread problem. However, these techniques become very complex and expensive for data rates greater than 20 Mbps.

A recent study given in [10] presents a feasibility analysis of indoor wireless communications at a very high data rate. A number of findings reported in the study are summarized as follows:

1. Neither equalization nor multicarriers can increase the data rate substantially above 20 Mbps.
2. The SNR can be increased by using a high-gain antenna array of M elements. The average gain in SNR is $10 \log M$. If antenna arrays are used in both basestation and remote stations, the average gain in SNR is $10 \log(M_b M_r)$, where M_b is the number of antenna elements in the basestation and M_r is the number of antenna elements in the remote station.
3. For a data rate greater than 20 Mbps, the coverage availability depends on the beamwidth. For example, for a 90% coverage availability at a 1-Gbps data rate, 5°beamwidth in both azimuth and elevation is required for coverage within a distance of about four rooms if an antenna array is used only in the basestation. In this case, 1650 antenna elements are needed for the two-dimensional array. Of course, this number is not very practical with respect to current technology.
4. Equalization does not significantly reduce the beamwidth requirement at very high data rates.
5. Antenna arrays should be used in both basestation and remote stations, thereby reducing the requirements for beamwidth and hence the number of antenna elements. For example, for a 90% coverage availability at a 1-Gbps data rate, 25°beamwidth is required for both the transmission and reception antennas. In this case, less than 50 antenna elements are needed for each array.
6. The beam direction for good reception (with small BER) cannot be determined *a priori*. In practice, one may have to find the beam direction that gives the best BER performance; that is, adaptive beamforming must be used

A study given in [11] presents an investigation into the use of adaptive beamforming in high-bit-rate indoor communications. The experimental system consists of a basestation and a remote station. Omnidirectional antenna elements are used n the basestation array. A single sectored patch antenna is used in the remote station. The operating frequency is 18.0 GHz and the channel bandwidth is 60 MHz. Using maximum-ratio combining, 15.9-dBm transmission power is required n a 100-ft office with a 30-ns rms delay spread to achieve the minimum 30 Mbps ata rate at a 10^{-4} uncoded BER using a four-element antenna. A six-element ntenna at a 10^{-2} outage rate can achieve 51 Mbps. However, only 6 dBm is required to achieve a similar performance (50 Mbps data rate at 10^{-5} BER) using daptive beamforming based on the LMS algorithm. The user identification codes nd a special training sequence are used for generating the reference signal. These esults demonstrate the effectiveness of adaptive beamforming in high-data-rate indoor communications. This fact is also supported by the results presented in a tudy given in [12], which show that the delay spread at the output of the adaptive ntenna can be significantly reduced. In particular, the rms delay spread in terms f bit duration is reduced from 0.7 with an omnidirectional antenna to 0.1 with a 6-element adaptive circular array.

9.2 WIRELESS MOBILE DATA COMMUNICATIONS

Wireless data services and systems represent a rapidly growing and increasingly mportant segment of the telecommunications industry. Mobile data communications systems, including both private and public networks, provide a wide variety of services: electronic mail, enhanced paging, modem and facsimile transmission, emote access to host computers and office local area networks (LAN), information broadcasting, and data services in an intelligent vehicle highway system. One nay anticipate that since the demand in a high system capacity is currently increasing rapidly, adaptive beamforming will play an important role in mobile data communications.

9.2.1 Wireless Data Networks

There are a number of existing private and public mobile data services networks, which include advanced radio data information system (ARDIS), MOBITEX, and cellular digital packet data (CDPD). Two other mobile data service standards are currently being planned: mobile data services based on IS-95 and the Trans-European Trunked Radio (TETRA).

ARDIS

ARDIS is a two-way radio service developed and implemented jointly by IBM and Motorola in 1983. The service is suitable for transfers of data files that are less than

10K. Remote users access the system from laptop radio terminals, which communicate with the basestations. The operating frequency band is 800 MHz and the transmit and receive frequencies are separated by 45 MHz. The RF channel data rate is 19.2 kbps, which provides a user data rate of 8 kbps. The portable units operate with 4W radiated power. The modulation technique is FSK, the access method is FDMA, and the transmission packet length is 256 bytes. The portables access the network using a random-access method called data sense multiple access (DSMA), which is closely related to the carrier sense multiple access/collision detection (CSMA/CD) access technique. A portable listens to the basestation transmitter to determine whether a busy bit is on or off. When the busy bit is off, the portable will transmit data packets. When two portables transmit at the same time, the data packet will collide. As a result, retransmission of these packets is required. The characteristics of the ARDIS network are listed in Table 9.3. Because of the use of overlapping-cell coverage in ARDIS, interference occurs when signals are transmitted simultaneously from adjacent basestations. The ARDIS network deals with this problem by turning off the adjacent transmitters, for 0.5 to 1 sec when an outbound transmission takes place. This scheme, of course, lowers the overall network capacity.

MOBITEX

The MOBITEX system is a nationwide interconnected trunked radio network, which was developed by Ericsson and Swedish Telecom in the mid-1980s. MOBITEX uses packet-switching techniques, as does ARDIS, to allow multiple users to access the same channel at the same time. The mobile units transmit at 896 to 901 MHz and the basestations at 935 to 940 MHz. The system uses dynamic power setting in the ranges of 100 mW to 10W for mobile units and 100 mW to 4W for portable units. The GMSK modulation technique is used, with $BT = 0.3$ and noncoherent demodulation. The transmission rate is 8 kbps, half-duplex is used in 12.5 kHz channels, and the service is suitable for file transfers up to 20K. In the MOBITEX system, a proprietary network layout protocol called MPAK is used, which provides a maximum packet size of 512 bytes and a 24-bit address field. The reservation-slotted ALOHA (R-S-ALOHA) random-access method is used in the system. The characteristics of the MOBITEX system are listed in Table 9.3.

CDPD

The CDPD system is designed to provide packet data services as an overlay onto the existing analog cellular telephone network, AMPS. The CDPD system employs the same 30-kHz channelization as that used in the existing AMPS system. Each

0-kHz CDPD channel will support channel transmission rates of up to 19.2 kbps.)egraded radio channel conditions, however, will limit the actual information pay-ɔad throughput rate to lower levels, typically 5 to 10 kbps, and will introduce dditional time delay due to the error-detection and retransmission protocols. The ;DPD radio link uses GMSK modulation, with $BT = 0.5$, at the standard cellular arrier frequencies for both uplink and downlink. The specific value of BT is cho-en so that the existing AMPS receiver can easily detect the CDPD format without edesign. In the CDPD system, the slotted-DSMA (S-DSMA) method is used. The haracteristics of the CDPD system are listed in Table 9.3.

'lanned Systems

'he IS-95 data services will generally be based on standard data protocols to the reatest extent possible. At the time of this writing, the standardization is still in rogress. The possible system characteristics of IS-95 data services are also listed in 'able 9.3. It should be pointed out that typical raw channel rates for digital cellular ransmission are measured at approximately a BER of 10^{-2}. However, the accept-ble data transmission usually requires a BER of about 10^{-6}. Thus, efficient use of RQ, error correction codes, and other effective means (e.g., adaptive beamforming) re required to deal with the error characteristics in the mobile environment.

TETRA is currently developed as a family of standards, which include a set of adio and network interface standards for trunked voice and data services, and an air nterface standard for both wide-area packet data services for both fixed and mobile sers and for supporting other standardized network access protocols. The proposed nodulation technique is $\frac{\pi}{4}$ differential quadrature phase-shift keying ($\frac{\pi}{4}$-DQPSK). 'he channel rate is 36 kbps in each 25-kHz channel. It has been reported that he TETRA standard is currently designed to accommodate two popular forms of nultiuser access, namely, the slotted ALOHA (with and without packet reservation) nd DSMA. The proposed system characteristics of the TETRA system are also sted in Table 9.3.

.2.2 Benefits and Performance

Aost of the benefits of using adaptive beamforming for land mobile communications, vhich were discussed in Section 8.2, are applicable to mobile data services. However, is worthwhile to mention a number of specific benefits:

1. Both the ARDIS and MOBITEX systems support portable communications from inside buildings. High levels of transmission power are required for both the portables and basestations to overcome the inbuilding signal penetration

Table 9.3

Characteristics and Parameters of Mobile Data Service Networks (After [13] and [9].)

	ARDIS	*MOBITEX*	*CDPD*	*IS-95 Data*	*TETRA*
Frequency (MHz)					
Downlink	800 with	935–940	869–894	869–894	400 & 900 bands
Uplink	45 MHz sep.	896–901	824–849	824–849	400 & 900 bands
Channel spacing	25 kHz	12.5 kHz	30 kHz	1.25 MHz	25 kHz
Channel access	FDMA	FDMA	FDMA	FDMA	FDMA
Multiuser access	DSMA	S-CSMA	S-DSMA/CD	CDMA-SS	DSMA & S-ALOHA
Modulation	FSK, 4-FSK	GMSK	GMSK	4-PSK/DSSS	$\frac{\pi}{4}$-DQPSK
Channel bit rate	19.2 kbps	8.0 kbps	19.2 kbps	9.6 kbps	36 kbps
Spectral efficiency	0.77	0.64	0.64		
Packet length	up to 256 bytes	up to 512 bytes	24–928 bytes		192 & 384 bytes
Carrier	private	private	public	public	public
Service coverage	major metro.	major metro.	all AMPS	all CDMA	trunked radio
Coverage type	indoor/mobile	indoor/mobile	mobile	mobile	mobile

problem. By using adaptive antennas, the high transmission power requirements can be eased to a large extent. Basically, the reduction in transmission power is about $10 \log M$ dB for both portables and basestations, where M is the number of antenna elements.

2. In the ARDIS network, the basestation transmitters are shut off temporarily to deal with interference between adjacent basestations, thereby decreasing the capacity of the network. Alternatively, the interference problem between adjacent basestations can be overcome by using adaptive antennas in the basestations. The adaptive antenna used in a basestation steers nulls, for both reception and transmission, in the direction of the adjacent basestations thereby protecting the basestation against being interfered with by others and preventing from interfering with others. This is accomplished without shutting down the basestations and therefore increases the capacity.
3. Using adaptive beamforming, the SNR at the output of the antenna array increases by $10 \log M$ dB on average in noise-limited environments, whereas in interference-limited environments, the additional improvement in SIR results from interference rejection. That is, the signal quality in a CDPD system or in an IS-95 data system can be improved by using adaptive beamforming.
4. With random-multiple-access techniques, such as the slotted ALOHA method only a single packet can be received successfully in a given slot. If there is more than one data packet within the same slot, collision will occur and both packets may not be received correctly. Packet collisions limit channel throughput to only 36% for slotted ALOHA. Using adaptive beamforming the number of packet collisions can be substantially reduced. Furthermore

the implementation of a multiple-beam adaptive antenna makes it possible for a basestation to receive multiple packets in a single slot. Consequently, the channel throughput can be increased greatly.

5. Although DSMA is used in some networks to reduce the probability of packet collisions, these collisions still occur because of propagation delay. Let us consider the following situation. At the beginning of a slot, the basestation transmitter indicates that it is idle by turning off the busy bit. A portable seizes the opportunity to transmit a packet. Due to propagation delay, it takes a certain time interval for this packet to arrive at the base station. Before this packet arrives, the busy bit is still off, thereby allowing other portables to transmit. When they do, packet collisions will certainly happen. Adaptive antennas can be used to deal with this problem (i.e., to reduce the probability of packet collisions).
6. In the CSMA scheme, a mobile unit monitors the channel by listening to others' transmission carriers. If the channel is idle, then the mobile unit will transmit a data packet. This approach can greatly reduce the probability of packet collisions. However, this scheme works effectively only if the mobiles are able to hear each other. However, in a practical radio environment, some mobiles may not be able to sense others' transmission. As a result, they may transmit packets, being convinced that the channel is idle when it is actually busy, thereby causing packet collisions. This problem is called hidden terminal problem [14]. The adverse effects of this problem can be greatly reduced by using adaptive beamforming.

In a study given by Ward and Compton in [15], the use of an adaptive antenna proposed to reduce packet collisions in a slotted-ALOHA packet radio network. s a result, the throughput can be greatly improved. Figure 9.6 shows the average iroughput as a function of the new-transmission probability with and without the se of adaptive beamforming. For a relatively small value of the new-transmission robability ($\sim$0.007), the system becomes unstable without adaptive beamforming. owever, with adaptive beamforming, the throughput keeps increasing to the maxiıum of near 0.83. Furthermore, the results show that the average delay experienced y a new packet is always better with adaptive beamforming than without. The nprovement in average delay is greater as the input traffic is increased.

In another study given by the same authors in [16], the use of a multiple-beam daptive antenna is proposed to capture multiple data packets in a single slot. 'hey have shown that when an eight-element adaptive antenna ($N + 1 = 8$) is sed, up to $K = 6$ packets can be captured during a single slot, thereby yielding a ıbstantially high level of throughput. Figure 9.7 shows that average throughput as function of the new-transmission probability. For relatively low traffic, the rise in

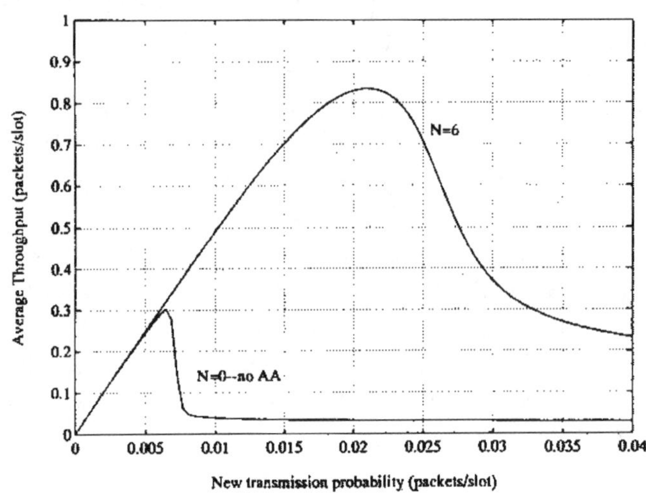

Figure 9.6 Average throughput as a function of the new-transmission probability wit and without the use of adaptive beamforming, where $N+1$ is the number c antenna elements. (From [15] ©1992 IEEE)

average throughput is approximately linear. Essentially, there is no backlog and th maximum average throughput is nearly achieved. The length of this linear regio increases with the number of beams K until the total traffic on average exceeds th capability of the adaptive antenna. Figure 9.8 shows the maximum throughput a a function of N, where $N+1$ is the number of antenna elements. It should be note that N is the number of nulls that can be formed per beam. $N=0$ corresponds t the standard S-ALOHA which has the maximum throughput of e^{-1} in theory. number of observations can be made from the figure:

1. For a given value of K, there is a minimum value of N above which no furthe improvement in throughput can be obtained.
2. As the value of K increases, the value of N should be increased in order t achieve the maximum throughput.
3. The improvement in throughput is not linearly related to K; that is, th increase in throughput becomes smaller with each additional beam.

Adaptive beamforming has also been proposed to deal with the hidden-termin problem in a slotted nonpersistent CSMA system, thereby improving its throughpu performance [17]. In the study, a single-hop packet radio system is considerec in which a very large number of users are distributed in the area covered by basestation. It is assumed that the time needed to detect the carrier of a packe

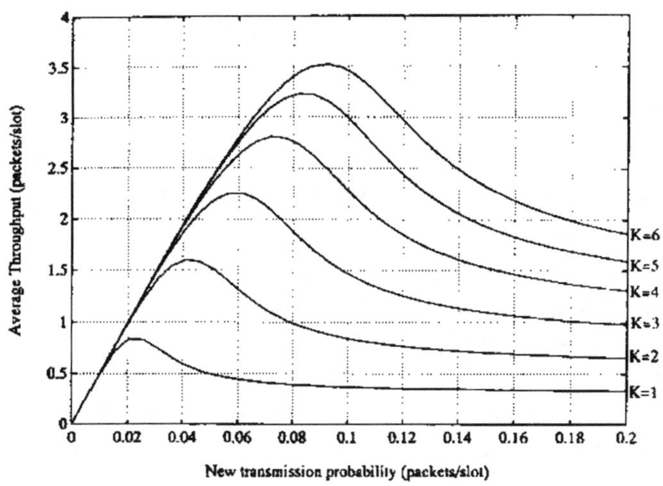

Figure 9.7 Average throughput as a function of the new-transmission probability for different number of beams. (From [16] ©1993 IEEE.)

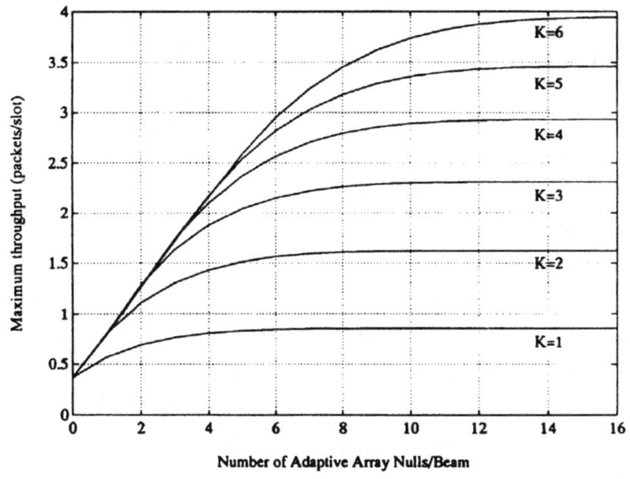

Figure 9.8 Maximum throughput with optimal weighting as a function of N and K. (From [16] ©1993 IEEE.)

and to process the packet is negligible. Each packet with a fixed width requires a certain time interval for transmission, and propagation delay is assumed equal to the slot size. The interarrival time between packets is mutually independent with an exponential distribution. The basestation acquires a single packet in a given slot. A number of findings have been obtained [17]:

1. In a high traffic condition, even without the hidden-terminal problem, the throughput with a four-element adaptive antenna is still better than that with a single antenna.
2. With the hidden-terminal problem, the throughput performance with a single antenna degrades greatly, whereas the throughput performance with the adaptive antenna degrades only slightly.
3. In addition to the improvement in throughput, adaptive beamforming can reduce the average number of retransmissions and schedulings with and without the hidden-terminal problem.

Caution should be exercised if one is to use these results for practice, because in this study, multipath has not been considered.

9.2.3 Acquisition

The major difficulty in the application of adaptive antennas to a wireless packet data system is the acquisition problem—the problem of forming a beam for the desired packet and suppressing the others. In practice, a data packet arrives the basestation at an unknown time and from an unknown direction. The adaptive antenna must form a desired beam in a very short period, which is a small fraction of the packet width, so that the system is in time to receive the message portion of the packet.

In order to deal with the acquisition problem, Ward and Compton proposed that a special preamble, called acquisition preamble, be added to the beginning of each packet [15]. The acquisition preamble is divided into two parts. The first part is to trigger the acquisition process, which is a 13-bit Baker code with a highly peaked aperiodic autocorrelation function. The second part is used as a reference for the adaptive antenna to form the desired beam, which is a PN code with multiple periods. Furthermore, the width of a time slot is made slightly larger than the packet width by an uncertainty interval. The packet transmission from each user is then randomized over this uncertainty interval in each slot, so that each packet arrives at a slightly different time. At the beginning of each time slot, when the basestation is ready for acquiring a new packet, the antenna weights will be set or reset to the quiescent pattern; that is, the antenna listens to the signals in all directions. When a packet arrives at the basestation, the output of the antenna beamformer will be passed through the circuit loop described in Section 3.5. The first part of the acquisition preamble is used to generate a peak pulse, which serves as a timing spike to trigger the generation of a reference signal using the second part of the acquisition preamble. Within the duration of the second part of the acquisition preamble, a number of iterations take place for the weights to converge to a desired state for an optimal reception of the message portion of the packet. At the end of the preamble, the converged weights are frozen for the rest of the packet.

The two-part acquisition preamble is suitable for real-time operation. A simpler acquisition preamble can also be used with additional hardware at the basestation. The acquisition preamble in each packet consists only of a number of periods of a single PN sequence [16], thereby reducing the packet width. However, this requires a data buffer to store the sampled data, as shown in Figure 9.9. When a packet arrives at the basestation, the acquisition preamble of this particular packet is correlated with a stored replica of the preamble PN sequence. If the correlation coefficient is above a certain threshold, weight adaptation begins and the converged weights will be fixed at the end of the preamble. While this takes place, the array continue to sample the incoming signals and store the data in the buffer. When the weights are ready, beamforming is then applied to the message portion of the packet.

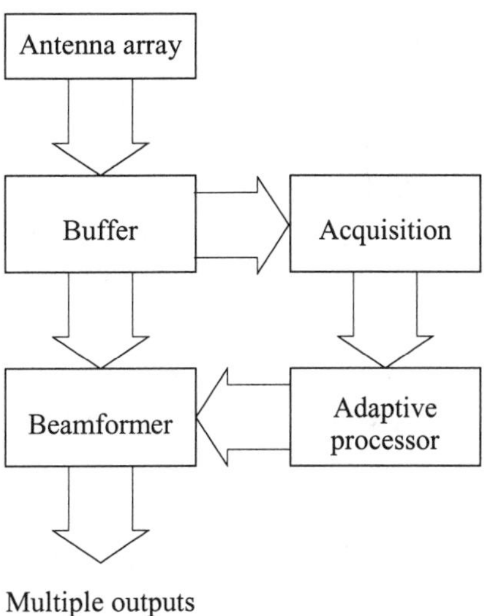

Figure 9.9 Adaptive beamforming configuration for packet radio communications.

To acquire multiple packets in a single slot, a signal processing scheme was proposed in [16]. Multiple beams are formed for packets arriving in a slot. A bank of threshold detectors are used to acquire multiple packets, as shown in Figure 9.10. When multiple packets arrive within the same slot, the first packet will trigger threshold detector 1 (TD1), which will instruct the adaptive processor to start the weight adaptation for the first packet. TD1, when triggered, will also enable TD2 immediately. When the second packet arrives, it will trigger TD2, which then instructs the processor to begin weight computation for the second packet. TD2 will also enable the next threshold detector. A similar process will take place for the next detector, and so on.

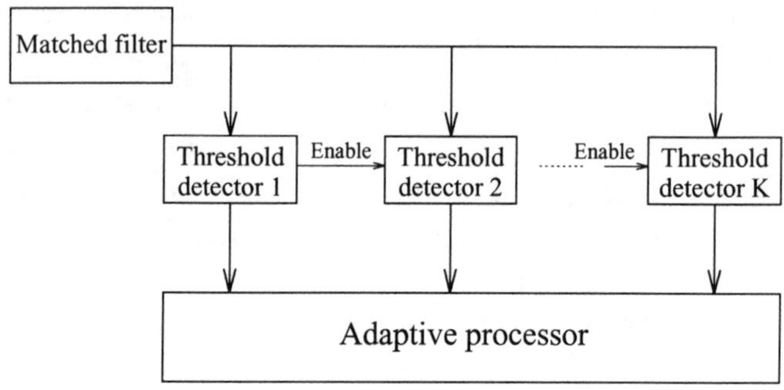

Figure 9.10 Acquisition signal processing scheme for multiple packets per slot.

Since all the data packets use the same preamble, careful timing is required for successful acquisition of multiple packets. Let us consider two data packets from two users arriving within the same slot. As long as packets arrive at least one-bit duration apart, they can be received successfully, provided that the following conditions are met:

1. Each threshold detector can only be triggered once within a given slot. Even though later packets arriving within the same slot cause additional timing spikes, they can only trigger the threshold detectors which have not been used.
2. The length of the uncertainty interval has to be set just less than one period of the preamble PN sequence. As a result, the latest packet arriving in a given slot will be present during the second period of the first packet preamble, when the weights are still being adapted. Thus, the interference from the other packets is taken into account as the weight adaptation progresses.
3. When packets are at least one-bit duration apart, their preambles will essentially be uncorrelated with each other. Therefore, they will have a minimal effect on each other's weight adaptation process.

When two packets arrive in a time interval of less than a one-bit duration, they collide with each other and the system will lose both packets. This is because all packets have the same preamble, and the system has no way to distinguish them when they arrive within a one-bit duration. Retransmission of these packets must be scheduled and carried out.

References

[1] H. Hashemi, "The indoor radio propagation channel," *IEEE Proc.*, vol. 81, pp. 943–968, July 1993.

[2] W. C. Y. Lee, *Mobile Communications Design Fundamentals*, John Wiley & Sons, New York, 1993.

[3] T. Lo and J. Litva, "Angles of arrival of indoor multipath," *Elect. Lett.*, vol. 28, Aug. 1992.

[4] J. H. Winters, "Optimum combining for indoor radio systems with multiple users," *IEEE Trans. Comm.*, vol. 35, pp. 1222–1230, Nov. 1987.

[5] S. A. Hanna, M. S. El-Tanany, S. A. Mahmoud, and S. R. Todd, "Applications of adaptive antenna combining to digital radio communications within buildings," in *Proc. 1992 IEEE Int. Conf. Selected Topics in Wireless Communications*, Vancouver, B. C., pp. 107–110, June 1992.

[6] P. Kwala and A. U. H. Shiekh, "Adaptive multiple-bem array for wireless communications," in *IEE Eighth Int. Conf. Antennas & Propagat.*, Heriot-Watt University, U.K., 1993.

[7] M. A. Bree, *Optimum combining with the LMS algorithm for indoor multipath channels*, PhD thesis, University of Saskatchewan, Saskatoon, Saskatchewan, 1994.

[8] A. A. M. Saleh and R. A. Valenzuela, "A statistical model for indoor radio propagation," *IEEE J. Sel. Areas Comm.*, vol. 5, pp. 128–137, Feb. 1987.

[9] T. S. Rappaport, *Wireless Communications: Principles and Practice*, Prentice Hall, Englewood Cliffs, NJ, 1996.

[10] M. J. Gans, R. A. Valenzuela, J. H. Winters, and M. J. Carloni, "High data rate indoor wireless communications using antenna arrays," in *Sixth IEEE Int. Symposium on Personal, Indoor and Mobile Radio Communications*, Toronto, pp. 1040–1046, Sept. 1995.

[11] S. Nam, I. D. Robertson, and A. H. Aghvami, "Study of optimum combining diversity for high bit rate indoor communication systems using sectored antenna," in *IEE Conf. Telecommunications*, pp. 337–341, 1990.

[12] C. Passerini, M. Missiroli, G. Riva, and M. Frullone, "Adaptive antenna arrays for reducing the delay spread indoor radio channels," *Elect. Lett.*, vol. 32, pp. 280–281, Feb. 1996.

[13] A. H. Levesque and K. Pahlavan, "Wireless data," in J. D. Gibson, ed., *The Mobile Communications Handbook*, CRC Press and IEEE Press, 1996.

[14] F. A. Tobagi and L. Kleinrock, "Packet switching in radio channels: Part II—The hidden terminal problem in carrier sensing multiple access and the busy tone solution," *IEEE Trans. Comm.*, vol. 23, pp. 1417–1433, Dec. 1975.

[15] J. Ward and R. T. Compton, "Improving the performance of a slotted ALOHA packet radio network with an adaptive array," *IEEE Trans. Comm.*, vol. 40, Feb. 1992.

16] J. Ward and R. T. Compton, "High throughput slotted ALOHA packet radio networks with adaptive arrays," *IEEE Trans. Comm.*, vol. 41, March 1993.

17] A. Sugihara, K. Enomoto, and I. Sasase, "Throughput performance of a slotted nonpersistent CSMA with an adaptive array," in *M. E.th IEEE Int. Symposium on Personal, Indoor and Mobile Radio Communications*, Toronto, pp. 633–637, Sept. 1995.

Chapter 10

Antenna Elements

The role of antennas in wireless communications systems is to establish a radio transmission line between radio stations, such as a basestation and a mobile station, and a communications satellite and a mobile station. Various types of antenna elements can be used for wireless communications, depending on the types of applications. This chapter is intended to give readers technical flavor of some common antenna elements that can be used in a DBF array. These antenna elements include linear antenna elements, horn antenna elements, and printed antenna elements. For more comprehensive descriptions of antennas, readers are referred to a number of antenna handbooks [1–3].

In general, antennas for wireless communications have been required to be small, light-weight, and low-profile, and to have a specified radiation pattern. In addition, antennas must exploit ambient propagation characteristics. With the revolution of wireless communications systems, antenna technology has also progressed, and the design concept has changed as well, although the fundamentals essentially remain the same. Requirements for antennas depend on the types of wireless communications systems. Among the commonly used antenna elements, monopole and dipole antennas are popular in terrestrial communications and horn antennas and helical antennas are considered to be suitable for satellite communications applications. With the advances in printed antenna technology, printed antennas are finding more and more applications in wireless communications. In future communications systems, either terrestrial or spaceborne, they will play an important role because of their desirable characteristics, such as light weight and small volume.

10.1 LINEAR ANTENNA ELEMENTS

Linear antennas have relatively simple structures. They include dipole and monopole antennas. Some examples of linear antenna elements are shown in Figure 10.1. The basic dipole antenna structure is a center-fed linear cylindrical antenna. When the length of each arm of the dipole approximately equals half a wavelength ($\frac{\lambda}{2}$), its input impedance becomes pure real, and impedance matching can be done using a transmission line with a characteristic impedance of 73.13Ω. It is also understood that the variation in impedance with frequency of an antenna of a larger diameter is much less than that of an antenna of a smaller diameter. Therefore, an antenna with a larger diameter supports more broadband operation. The radiation pattern of a dipole has a shape like a donut, as shown in Figure 10.2. The cross-section on the vertical plane appears to be two symmetrical lobes, whereas on the horizontal plane, it is omnidirectional. It should be pointed out, though, that the principal dipole antenna is practically never used in vertical polarization applications, due to the problem of mounting and feeding while maintaining symmetry.

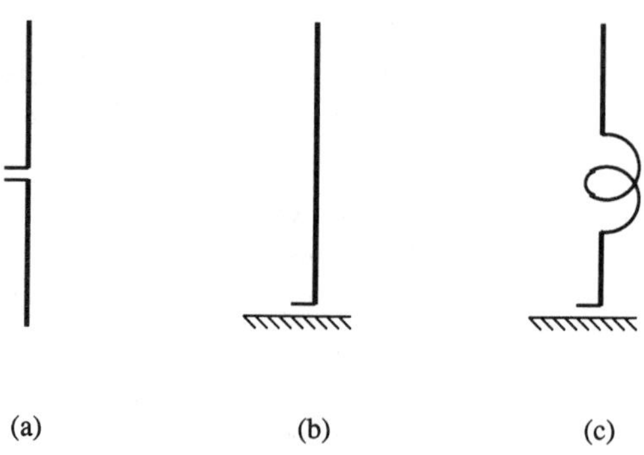

Figure 10.1 Linear antenna elements: (a) dipole antenna; (b) monopole antenna; (c) monopole antenna with a load.

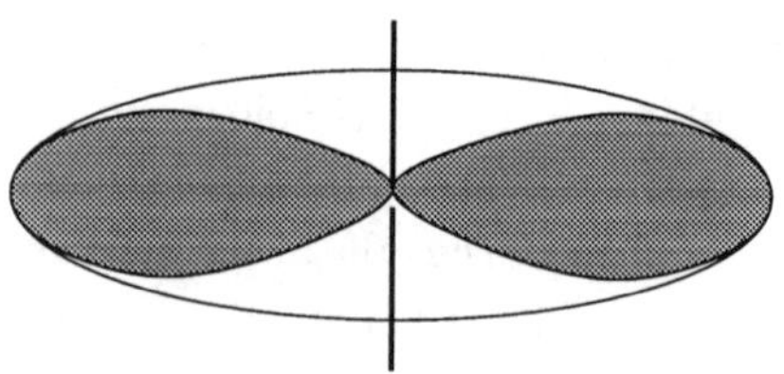

Figure 10.2 Radiation patter of a dipole, which has the shape of a donut.

Monopole antennas are the most used antennas in mobile communications, but have the simplest structure. The radiation element is mounted on a ground plane. If the ground plane were infinitely large and were a perfect conductor, the radiation pattern and bandwidth characteristics of the antenna would be the same as those of a dipole antenna, due to the effect of an image of the element formed by the ground plane. However, the input impedance is only half that of the dipole antenna (e.g., 36Ω for a quarter-wavelength monopole antenna). Theoretically, the directivity is 3 dB larger than that of the dipole, because the radiation power is radiated only to the upper half space of the ground plane. However, because of the physical size and the conductive loss of the real-world ground planes, the practical directivity cannot be improved by this amount. In actual applications, the size of the ground plane is finite, and the direction of maximum radiation tilts somewhat upwards from the horizontal plane. Therefore, the effective gain of this antenna is usually lower than that of a dipole antenna. The length of a commonly used monopole antenna can be in the range of 1/4-wavelength to 5/8-wavelength. A 1/4-wavelength monopole antenna is the fundamental mobile antenna. A 5/8-wavelength monopole antenna has comparatively large directivity in the horizontal direction. Thus, a 5/8-wavelength monopole antenna is often used as a high-gain antenna in combination with a ground plane. The reason why the gain is higher than that of a 1/4-wavelength monopole antenna is that its antenna aperture is larger, and the rise radiation beam due to the finite ground plane is less than that of a 1/4-wavelength monopole antenna. The antenna's input impedance is roughly 50Ω.

The omnidirectional pattern on the horizontal plane makes the monopole antenna elements popular in mobile communications. This property is not desirable in array applications. It is easy to see the problem in the linear array. To illustrate this, we refer readers back to the array factor of a linear array given by (2.11). It is obvious that $F(\theta) = F(180° - \theta)$; that is, the array factor is not a unique function in the range $[-\pi, \pi]$. This ambiguity problem is overcome in practice by using direction antenna elements. In the case of a linear array of dipole or monopole antennas, a reflector must be used on one side of the linear array. The ambiguity problem does not exist in the case of a circular array. One may expect intuitively that it is desirable to use omnidirectional elements in a circular array. However, in practical applications, this is not the case. In the circular array shown in Figure 2.3, the desired array pattern can be broken down into a series of complex Fourier spatial harmonics; that is,

$$F(\phi) = \sum_{n=-N}^{N} C_n j^n J_n(\kappa R) e^{jn\phi} \tag{10.1}$$

where C_n represents the complex Fourier coefficient for the nth spatial harmonic and $J_n(\cdot)$ denotes the Bessel function. The Bessel function may go to zero for certain value of κR. This means that the array pattern is not stable as frequency changes over the band of interest. In some cases, the change in the pattern can be significant even if the change in frequency is small; that is, the bandwidth of the array is very narrow. This may pose a severe problem in the use of the circular array in wireless communications, especially in wideband applications. The array pattern can be stabilized using directional antenna elements, such as horns and printed antennas.

10.2 HORN ANTENNA ELEMENTS

A horn antenna can be viewed as a flared-out (or opened-out) waveguide. The horn produces a uniform phase front with a larger aperture than the waveguide and thus has a greater directivity [4]. Several types of horn antennas are illustrated in Figure 10.3.

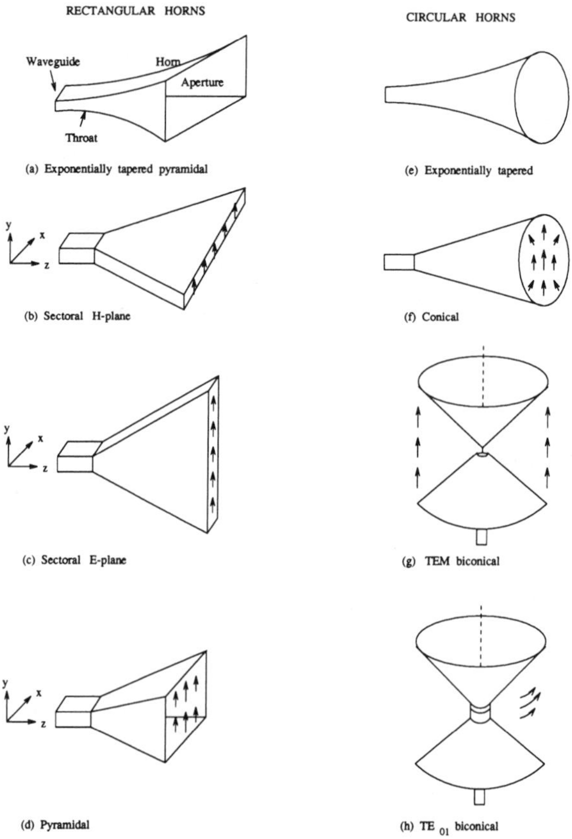

Figure 10.3 Types of rectangular and circular horn antennas.

The conventional conical, smooth-walled horn antenna has the simplest structure in the horn-antenna-type family but does not have the best polarization purity or even the highest gain [5]. Multimode and hybrid-mode horns radiate axially symmetric patterns with extremely low cross-polarization. Thus, they closely approximate the ideal Huygens source and, for this reason, are widely used as feeds for reflector antennas [6].

10.2.1 Multimode Horns

The lack of axial pattern symmetry in large conical horns may be overcome by introducing the TM_{11} mode along with the dominant TE_{11} mode in the so-called dual-mode conical horn, or Potter horn (Figure 10.4) [6]. The TM_{11} mode is generated at the step where the horn radius changes abruptly from a_0 to a_1 and is correctly phased at the aperture by proper choice of the length l of the straight section that follows the step (see Figure 10.4). The presence of the TM_{11} mode has essentially no effect on the H-plane aperture distribution of the horn, nor on its H-plane radiation pattern. But when properly phased and combined with the TE_{11} mode, it exerts a profound effect on the E-plane aperture distribution and the corresponding radiation pattern. Since the TM_{11} mode does not radiate axially, this device has less axial gain than a dominant-mode conical horn with the same aperture size. Its use, therefore, is not dictated in applications requiring maximum aperture efficiency. However, most horn applications are more concerned with pattern circularity and sidelobe performance. The Potter horn is thus a very satisfactory feed in a Cassegrain system, for example. Similar results can be realized in a square aperture pyramidal horn by means of a step in the throat region to cause conversion of dominant TE_{10} mode energy to a hybrid mixture of TE_{12} and TM_{12} modes [6].

For large reflector antennas, the use of the Potter horn has been largely superseded by the wider bandwidth hybrid-mode corrugated horn, but when the weight is critical and required bandwidth small, the dual-mode smooth-walled horn still finds applications. In satellite antennas, the dual-mode horn is used more frequently because of its simplicity and lower cost [5]. The bandwidth limitation is the most critical restriction of the dual-mode horn.

A good example of analysis and design of a Potter horn by the modal-matching technique with optimization of the horn's geometry through the minimax algorithm is given in [7].

10.2.2 Soft and Hard Horns

Lier and Kildal [8] have introduced the terminology of soft and hard horns to distinguish between horn antennas with soft and hard boundaries related mostly to the context of hybrid-mode horns. A soft boundary gives zero field intensity at the wall.

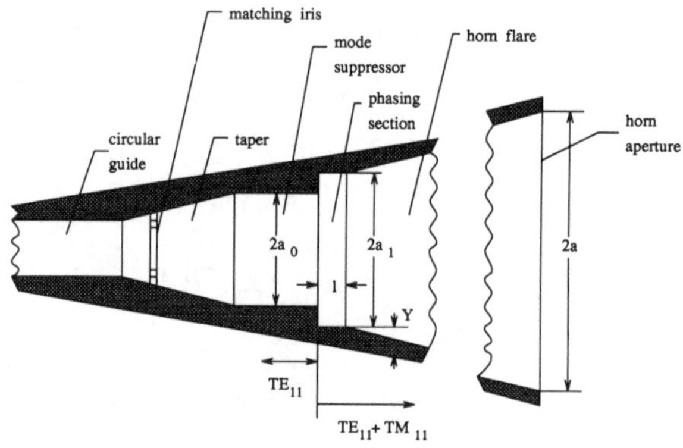

Figure 10.4 Dual-mode conical, or Potter, horn.

Classical hybrid-mode horns have a soft boundary. A hard boundary corresponds to a uniform field distribution over the horn aperture. In other words, soft and hard can be recognized as Dirichlet and Neumann types of boundary conditions, respectively.

Hard boundary conditions give higher directivity and smaller beamwidth due to the increased efficiency compared to soft conditions. The disadvantage is that the sidelobe level is much higher. The hard horn has significantly higher directivity than a soft horn only if it has almost constant phase over the aperture. Such a horn is also termed narrowband, since it has a frequency-dependent beamwidth.

10.2.3 Corrugated Horns

Corrugated horns (Figure 10.5) are hybrid-mode horns. These horns were developed to improve antenna efficiency and provide reduced spillover in large reflector antennas [6]. Corrugated horns can provide reduced edge diffraction, improved pattern symmetry and very low levels of cross-polarization. The very low levels of cross-polarization radiated by corrugated horns are essential for dual-polarization operation or frequency reuse. Compared to smooth-walled circular horns that have peak cross-polar levels of about -19 dB, the impedance properties of the corrugated horn push this peak level at the design frequency well below -30 dB, the actual peak level being limited by secondary design and manufacturing factors [9].

The effect of the circumferential slots or corrugations is to force H_ϕ to vanish along with E_ϕ at the horn wall. As a result, the boundary conditions for TE and TM waves are identical and the natural propagating modes are linear combinations

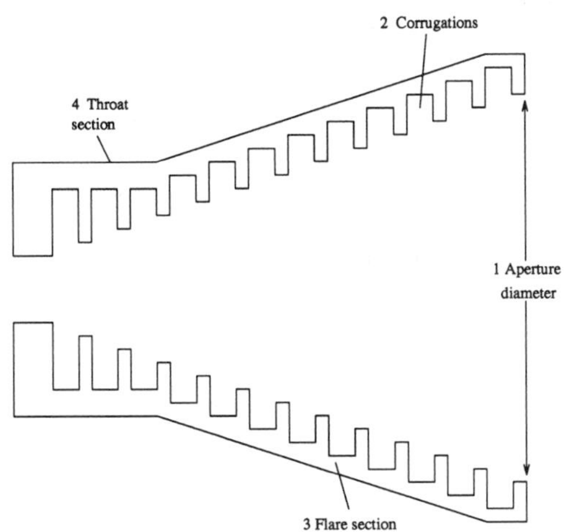

Figure 10.5 Corrugated horn.

of the two, and are referred to as hybrid modes [6]. Since corrugated horns are characterized by zero field intensity at the horn wall, which gives a low radiated sidelobe level, they can be referred to as soft horns [8]. Additionally, since the very low cross-polarization means that the fields in the aperture are essentially scalar, the corrugated horns are also called *scalar* horns, but this property has most significance for long horns or for wide-flare-angle horns.

Corrugated horns do not have the bandwidth limitation of the dual-mode horns. The bandwidth of corrugated horns is quite large and is not usually a limiting factor in their use. Corrugated horns can be analyzed using several methods ranging from a simple first-order qualitative assessment of performance to the very accurate modal-matching technique, which is also computationally intensive [9].

Corrugated horns have to be accurately made in order to work in the desired manner; therefore, it is an inherently expensive horn to produce, partly because of the corrugated surface and partly because it should be precisely fabricated in order to produce the potentially high quality radiation pattern. There are two main methods for manufacturing corrugated horns. One method is to machine them on a high-quality, numerically controlled lathe. The other method is electroforming. This second method is a particularly useful technique at millimeter waves. Another disadvantage of corrugated horns is their relatively high mass and volume, though most microwave horns are made out of aluminum in order to save mass.

The conical corrugated horns are by far the most common type. It is also possible to make rectangular and elliptical cross-section corrugated horns. Rectangular and elliptical horns have the attribute of different beamwidths in orthogonal planes as well as potentially low cross-polarization. However, they are difficult to design accurately and very expensive to manufacture, so they are not often used. A preferred solution when unequal beamwidths are required from a reflector antenna would be to use a conical corrugated horn and a shaped reflector.

We can distinguish three versions of the corrugated horn [9]:

1. Narrow-flare-angle corrugated horn;
2. Wide-flare-angle corrugated horn;
3. Compact or profiled corrugated horn.

The design of the compact or profiled corrugated horn requires extra care and is more sensitive to the geometry along the horn, but it has a number of advantages:

1. The horn can be about two-thirds as long as the equivalent constant-flare-angle horn while maintaining the low level of cross-polarization.
2. The flare angle at the aperture is zero, so the horn radiates as an open-ended waveguide, which gives the highest gain and efficiency for a given diameter.
3. The phase center, or point at which the energy radiates, is near the aperture and does not move as the frequency changes.
4. The small flare angle at the throat is ideal for achieving a good impedance match to the input waveguide.

10.2.4 Dielectric-Loaded Horns

Dielectric-loaded horns are also hybrid-mode type horns. There are several types of horns that fall into the category of dielectric-loaded horns. Common to all these is the fact that the interior of the horn is partly loaded with a dielectric material. Among the various dielectric-loaded horns we can distinguish:

1. Dielectric core (dielcore) horns comprising a dielectric core inside the horn separated from the metal wall by a dielectric layer (which may be air) with a lower dielectric permittivity than the core material [10–14].
2. Horns with a lossy dielectric lining on the horn wall [15,16].
3. Strip-loaded horns using circumferentially or transversely oriented conducting strips on dielectrically lined smooth-walled horns [17–19].
4. Horns with corrugated dielectric lining on the wall [20].

The dielectric core (Figure 10.6) horn has similar electrical properties as those of the corrugated horn. The advantages of dielcore horns compared to corrugated horns are:

1. Simpler design (lower price);

2. Simpler design criteria;
3. Low cross-polarization over a wider bandwidth than the equivalent corrugated horn;
4. Achieving the desired phase front by shaping the front surface of the core.

The disadvantages of the dielectric core horns are:

1. Dielectric gain loss;
2. Higher effective noise temperature;
3. Lower maximum available power.

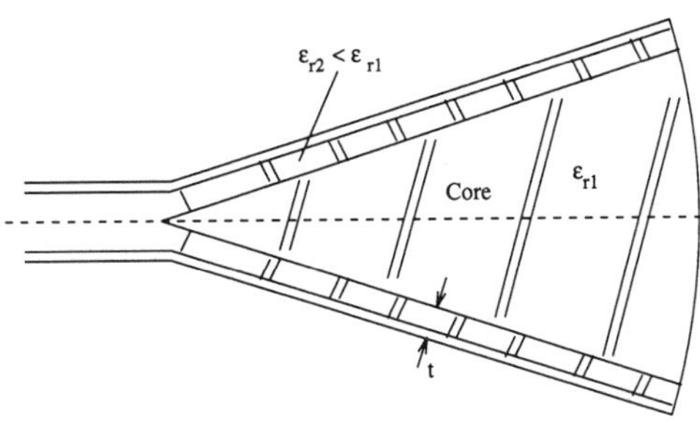

Figure 10.6 Dielectric core (dielcore) horn.

The dielectric core horns have been analyzed using the plane wall and the circular cylindrical waveguide model [10] and the modal-matching technique [11]. They are an interesting alternative to existing horn antennas. They have a very simple design and it may also be applied in antenna systems where hybrid-mode horns are not normally used due to their high prices [10]. More effort is required to develop space-qualified dielectric material with low weight and sufficient uniformity, and to develop low-loss dielectric material for high-power applications [14].

Horns with a lossy dielectric lining on the horn wall have characteristics similar to those of the corrugated horns but over a larger bandwidth [15]. Dielectric loading by dielectric lining on the horn wall has also been combined with the technique of multimode generation of higher mode by a step discontinuity to improve the aperture efficiency of pyramidal and sectoral horns [21, 22]. These horns have high aperture efficiency (on-axis gain), low sidelobe levels, and low cross-polar radiation, similar to corrugated horns.

The strip-loaded horn (Figure 10.7) consists of a hollow dielectric waveguide whose outer surface is coated with circumferentially oriented conducting strips with a periodic variation along the horn [17]. This horn exhibits a low sidelobe radiation

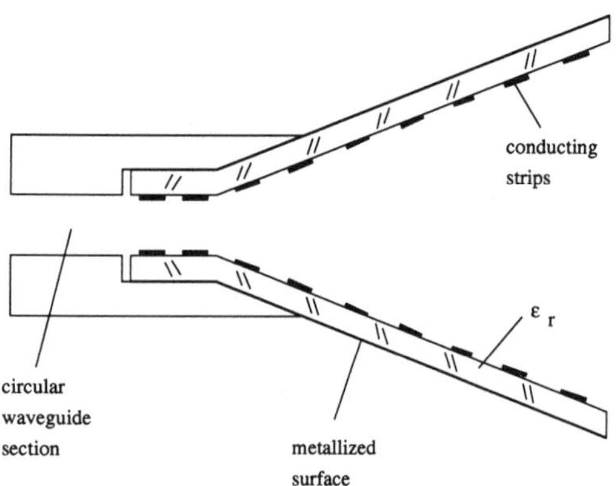

Figure 10.7 Circumferentially strip-loaded horn.

pattern similar to that of the corrugated horn over an even larger bandwidth and a low cross-polar bandwidth.

Limitations of this type of horn are mainly due to the lack of proper dielectric core material with sufficiently low losses and sufficient homogeneity. This horn has the potential of very low weight. It may also be cheaper to manufacture than the corrugated horn by using a combination of photolithography and etching processes. A rectangular strip-loaded horn was developed using such techniques [19] and was called a *simulated corrugated hor*, since conducting strip gratings with a metallic ground plane on a dielectric substrate are similar to metallic corrugated plates. The advantages of the strip-loaded horn are:

1. Possible substitute for a metallic corrugated horn;
2. High gain;
3. Low cross-polarization;
4. Low sidelobes;
5. Low cost;
6. Light weight;
7. Possibility of mass production with simpler methods.

The circumferentially strip-loaded horn can be referred to as a soft horn. Hard horns can be realized by corrugating the conducting wall of a smooth horn with longitudinal rectangular grooves filled with dielectric material or, in another way, by employing longitudinal conducting strips on a dielectric-loaded smooth conducting wall, the second configuration being cheaper to manufacture and lighter. Ideal hard horns are of little interest for practical applications because of the unwanted cross-polar radiation from the dielectric wall region and limited bandwidth. These limitations have been considered in the modified hard horn, which has a slightly tapered aperture distribution.

An array of hard horns would yield high directivity for limited scan applications and reduced spillover loss in cluster feeds. Lier et al. [14] even suggest that hard horn elements might be considered in the mobile link space antenna of the future land mobile satellite system in Europe. But hard horns are not yet well characterized. They need to be studied in more detail to see if the expected performance can be obtained in practice [23]. For example, the bandwidth potential of hard horns would require further investigation, particularly the excitation of higher order modes [14].

Circumferential corrugations at regular interval of the dielectric lining on the horn wall are said to be able to sustain a lower cross-polar level over a wider bandwidth relative to the continuously, dielectrically lined horn [20].

10.3 PRINTED ANTENNA ELEMENTS

When the number of array elements is large, the printed element approach is a natural choice [24], but the increased length and complexity of the feed structure reduces the advantage of the printed array concept substantially and makes it less interesting unless distributed transmit/receive (T/R) modules are used [25].

The generic microstrip patch is an area of metallization supported above a ground plane and fed against the ground at an appropriate point or points, as shown in Figure 10.8 [3]. The two most common shapes are the rectangular patch and the circular patch. The antenna element radiates a relatively broad beam broadside to the plane of the substrate. It has a very low profile so it can be made conformable. It can also be fabricated using printed circuit (photolithographic) techniques and hence at potentially low cost. Other advantages include simplicity, light weight, easy fabrication into linear or planar arrays, and easy integration with microwave integrated circuits (MIC) [25]. Easy integration with MICs means that printed antenna technology allows a great level of integration with feed and control circuitry, leading to a monolithic array antenna.

The original printed antenna configuration had some disadvantages: narrow bandwidth, spurious feed radiation, poor polarization purity, limited power capacity, and tolerance problems. Much of the development work in printed antennas has gone into trying to overcome these problems in order to satisfy increasingly stringent system requirements.

Microstrip antennas operate best when the substrate is electrically thick with a low dielectric constant. On the other hand, a thin substrate with a high dielectric constant is preferred for microstrip transmission lines and microwave circuitry. Some

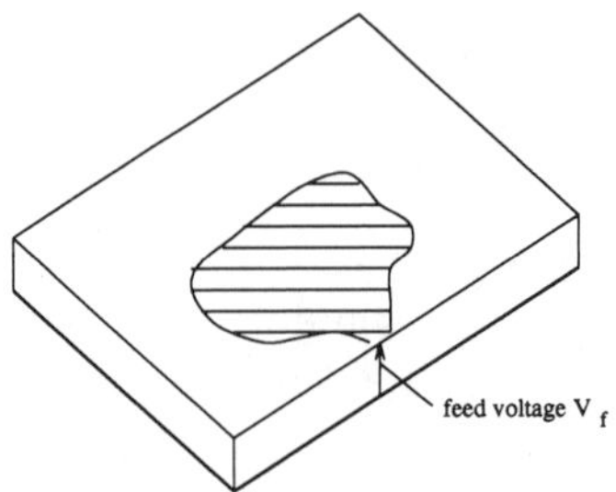

Figure 10.8 Generic microstrip patch antenna.

compromise has thus to be made between good antenna performance and good circuit performance [25].

Losses in a microstrip antenna occur in three ways [25]: (1) conductor loss, (2) dielectric loss, and (3) surface wave excitation. Except for extremely thin substrates, conductor and dielectric losses for a microstrip element are quite small, usually accounting for no more than a few percent loss in radiation efficiency. Surface waves bound to the dielectric substrate can be excited by the antenna, and since it does not contribute to the primary radiation pattern of the antenna, surface wave power is generally considered as a loss mechanism. Surface wave can also diffract from substrate edges or other discontinuities to degrade the antenna radiation pattern or polarization characteristics. Surface wave power generated by a single element increases with substrate thickness and is another reason for preferring a substrate with a low dielectric constant. Besides these inherent antenna element losses, the effect of losses due to feed lines and associated feed circuitry must also be considered often. These can be the dominating loss contributions in a large array.

10.3.1 Feed Methods

There are two groups of feeding structures for microstrip antennas, which can be further divided into subgroups [25]:

1. Direct contacting feed:
   (a) Microstrip line feed;
   (b) Coaxial probe feed.
2. Noncontacting feed:
   (a) Proximity feed (often referred to as *electromagnetic couplin*);
   (b) Aperture coupling.

The two direct feeding method structures shown in Figure 10.9 are very similar in operation and offer essentially one degree of freedom (for a fixed patch size and substrate) in the design of the antenna element through the positioning of the feed point to adjust the input impedance level. Direct contacting feed is relatively simple, but it has several disadvantages. A direct-fed printed antenna suffers from a bandwidth/feed radiation trade-off in which an increase in substrate thickness for the purpose of increasing bandwidth leads to an increase in spurious feed radiation, increased surface wave power, and possibly increased feed inductance. In practice, these antenna elements are limited in bandwidth to about 2% to 5%. In a coaxial probe feed, the patch element is fed through the ground plane from a parallel feed substrate, but in an array having thousands of elements such a large number of solder joints makes fabrication difficult and lowers reliability (an especially important consideration in space applications). Finally, although probe and microstrip line feeds excite the dominant mode of the patch element, the inherent asymmetry of these feeds generates some higher-order modes that produce cross-polarized radiation.

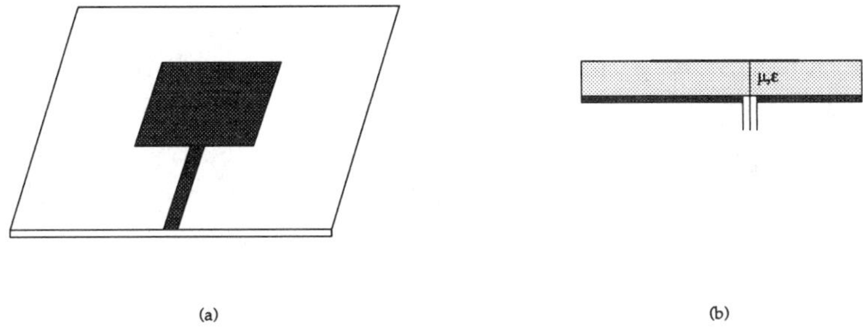

Figure 10.9 Direct feeding structures: (a) microstrip line; (b) coaxial probe.

The proximity feed structure, shown in Figure 10.10(a), consists of two substrate layers. A microstrip line is printed on the lower layer, which is terminated in an open stub below the patch, which is printed on the upper substrate. Proximity coupling has the advantage of allowing the patch to exist on a relatively thick substrate for improved bandwidth, while the feed line sees an effectively thinner substrate, which reduces spurious radiation and coupling. Fabrication is a bit more difficult because of the requirement for reasonably accurate alignment between substrates, but there is no soldering. The proximity-coupled patch has at least two degrees of freedom: the length of the feeding stub and the patch-width-to-line-width ratio. 13% bandwidth has been achieved with this configuration.

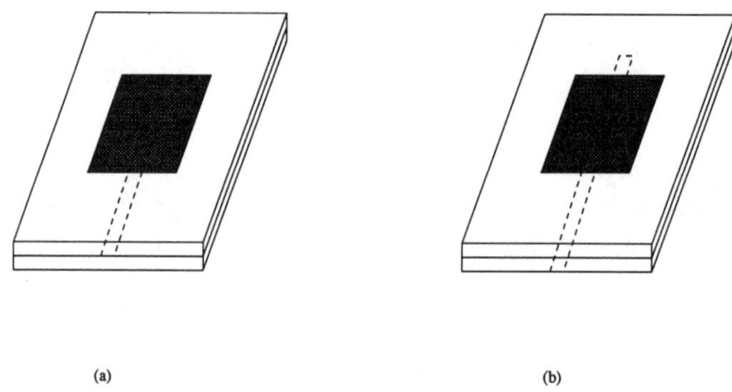

Figure 10.10 Noncontacting feeding structures: (a) proximity feeding; (b) aperture coupling.

The other type of noncontacting feed, aperture coupling, uses two parallel substrates separated by a ground plane as shown in Figure 10.10(b). A microstrip feed line on the bottom substrate is coupled through a small aperture (typically a narrow rectangular slot) in the ground plane to a microstrip patch on the top substrate. This arrangement allows a thin, high dielectric constant substrate to be used for the feed, and a thick, low dielectric constant for the antenna element, thus allowing impedance optimization of both the feed and the radiation functions. In addition, the ground plane prevents spurious radiation from the feed from interfering with the antenna pattern or polarization purity.

The geometry of the aperture-coupled patch has at least four degrees of freedom: the slot size, its position, the feed substrate parameters, and the feed line width. Impedance matching is performed by adjusting the size (length) of the coupling slot together with the width of the feed line, which is usually terminated in an open-circuited tuning stub. The maximum coupling occurs when the aperture is centered below the patch. In this case the excitation of the patch is symmetric, and this reduces the excitation of higher-order modes and leads to very good polarization purity. The aperture-coupled patch with a centered feed has no cross-polarization in the principle planes.

In all the above feeding structures, there is only a single feed point to generate linear polarization. It is possible to produce circular polarization with a single feed, but such techniques are usually very narrowband. It is more common to generate circular polarization by using two feed points to excite two orthogonal modes on the patch, with a 90° phase difference between their excitations (Figure 10.11(a)). This technique can be used with all the above feeding methods. An alternative technique used for small subarrays is to sequentially phase the excitations of an array of patches that radiate orthogonal sets of linear (or circular) polarization (Figure 10.11(b)).

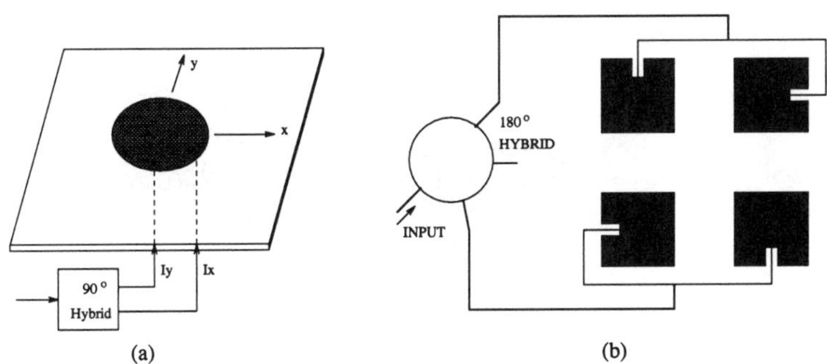

Figure 10.11 Techniques for obtaining circular polarization: (a) using two feed points for exciting two orthogonal modes on the patch with a 90° phase difference; (b) using a subarray of orthogonally polarized elements with sequential phase rotations (the phasings at the four patches are 0°, 90°, 180°, and 270°).

10.3.2 Bandwidth Improvements

The techniques used to overcome the narrow bandwidth (2% to 5%) of the traditional microstrip antenna element can be classified as either using an impedance-matching network or parasitic elements [25]. In both cases a double-tuning effect is often exploited.

A direct method of increasing the bandwidth of the microstrip element is to use a thick, low dielectric constant substrate. This inevitably leads to unacceptable spurious feed radiation, surface wave generation, or feed inductance. The bandwidth of the element is usually dominated by the impedance variation (the pattern bandwidth is generally much better than the impedance bandwidth). Thus, it is often possible to design a planar impedance-matching network to increase bandwidth. Up to 15% bandwidth has been obtained in this manner for probe-fed and microstrip line-fed elements. A bandwidth of 13% was achieved for a proximity-coupled patch element with a stub tuning network. If the matching network is coplanar with the antenna element, spurious radiation from the matching network may be a concern.

By using more complicated element geometry, increased bandwidth can be obtained by using parasitically coupled elements to produce a double-tuned resonance. One of the best ways to do this is with two stacked patches (Figure 10.12). The top patch is proximity-coupled to the bottom patch, which can be fed by one of th methods discussed earlier. 10% to 20% bandwidth has been achieved with probe-fe stacked patches, and 18% to 23% bandwidth has been achieved for aperture couple stacked patches.

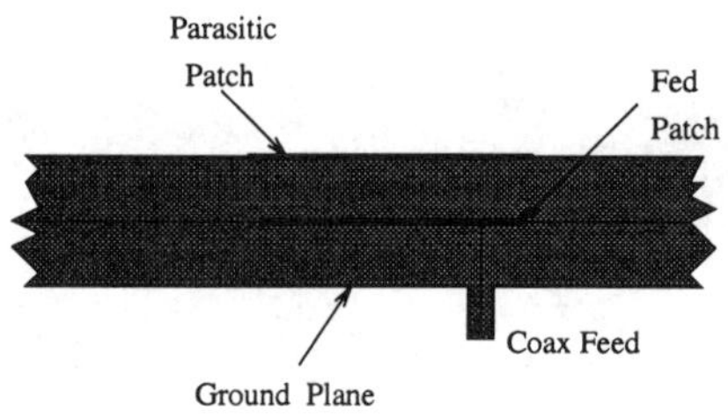

Figure 10.12 Bandwidth enhancement using stacked patches.

Double tuning can also be obtained with a single aperture-coupled microstri patch by lengthening the coupling slot so that it is near resonance. This simplifie construction relative to using stacked patches, but the larger coupling slot ma result in a higher back lobe level. Bandwidths in excess of 20% have been achieve for single elements and arrays using this technique.

It is also possible to use parasitic elements coplanar with a driven element t produce double tuning. This configuration can be a disadvantage in array applications because the lateral extent of such an element usually requires array spacing greater than $\lambda/2$. In addition, the phase center of this element will not be fixe over its operating bandwidth.

Other than using the bandwidth improvement techniques described above, ther are some printed antenna elements that provide a very wide bandwidth [26]. Fo example, the printed circuit balun-fed dipole antenna, which is commonly used i brick-type array constructions, operates over up to 40% bandwidth with good scanning characteristics (Figure 10.13). Other examples are the flared notch and "Vivaldi" elements which can provide over two octaves of bandwidth (Figure 10.14(a) and (b), respectively).

10.3.3 Power-Handling Capability

The power-handling capability of microstrip antenna elements for transmit applications is seldom mentioned in any references on such antenna elements. Here, we extend the considerations of the power-handling capability of microstrip lines [27,28] to the microstrip antenna elements, since they share the same technology.

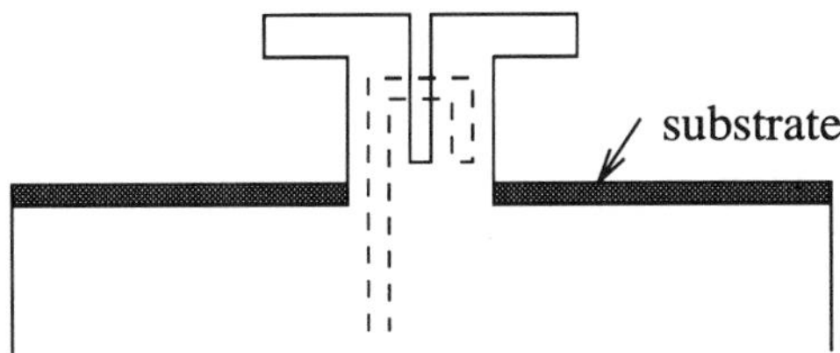

Figure 10.13 Dipole/balun radiating element.

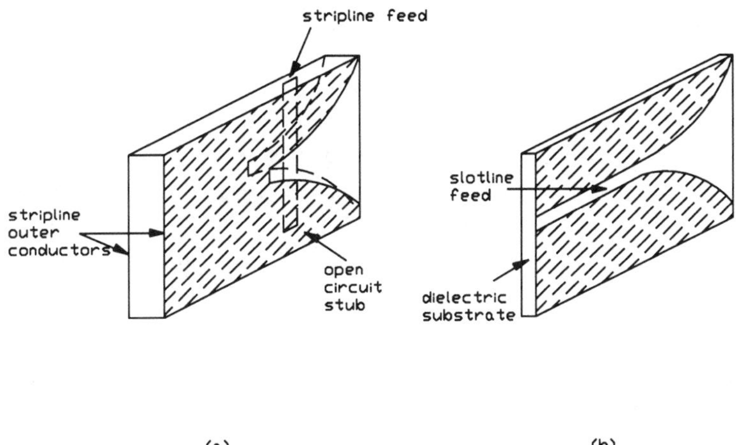

Figure 10.14 Printed antenna elements: (a) flared notch strip element; (b) flared "Vivaldi" slotline element.

Microstrip antenna elements (or microstrip lines) are not as well suited to high power applications such as horns (or waveguides), but they can be used for medium-power applications. As an example, a 50Ω microstrip line on a thin polystyrene substrate can handle a few hundred watts at 2 GHz. The power-handling capability of a microstrip antenna element, like that of a microstrip line or any other dielectric-filled transmission line, is limited by heating caused because of ohmic and dielectric losses and by dielectric breakdown. Increase in temperature due to conductor and dielectric losses limits the average power of the microstrip element, while the breakdown between the conductor patch and ground plane limits the peak power.

The average power-handling capability of microstrip elements is determined by the temperature rise of the conductor patch and feed and the supporting substrate.

The parameters that play major roles in the determination of average power capability are:

1. Conductor losses;
2. Thermal conductivity of the substrate material;
3. Surface area of the conductor patch and feed;
4. Ambient temperature (temperature of the medium surrounding the microstrip element).

Therefore, a dielectric substrate with low loss tangent and large thermal conductivity will increase the average power-handling capability of microstrip elements. Also the maximum average power-handling capability of microstrip elements decreases with frequency.

A thick layer of substrate can support higher voltages (for the same breakdown field). Hence, microstrip elements on thick substrates have higher peak power handling capability. The sharp edges of the conductor serve as field concentrators. The electric field tends to a large value at the sharp edges of the conductor if it is flat and decreases as the edge of the conductor is rounded off more and more. Therefore a thick and round conductor will increase breakdown voltage.

10.3.4 Examples

A printed sleeve antenna, as shown in Figure 10.15, was designed and fabricated for mobile communications applications [29]. This antenna can be used as an element in a DBF antenna array for a basestation. The length of this sleeve antenna is similar to that of a dipole antenna, but with the matching at the end rather than at the center. The substrate used for the antenna has a dielectric constant $\epsilon_r = 3.0$, with the dimensions of $W \times L \times H = 0.08\lambda \times 0.4\lambda \times 0.01\lambda$. The antenna was designed to work at 900 Mhz. The gain of the antenna is 5.2 dBi with a 25% input impedance bandwidth. The radiation pattern is maximum in the elevation angle of $30°$ to $70°$.

A lightweight, dual circularly polarized microstrip antenna for use in a communications satellite array-fed reflector was designed and fabricated [30]. The radiator is intended as a one-to-one replacement for existing waveguide feeds and is designed to operate at C-band. It consists of a four-element patch array fed by two microstrip polarizing networks. The radiator is considerably lighter and more compact than waveguide feeds having similar gain and radiation patterns. The entire assembly is space-qualified and designed to withstand launch conditions. The radiating portion of the compact radiator is composed of electromagnetically coupled patch elements. The measured on-axis axial ratio for the right-hand and left-hand circular polarizations that can be achieved from this radiator is about 0.3 dB across the band of

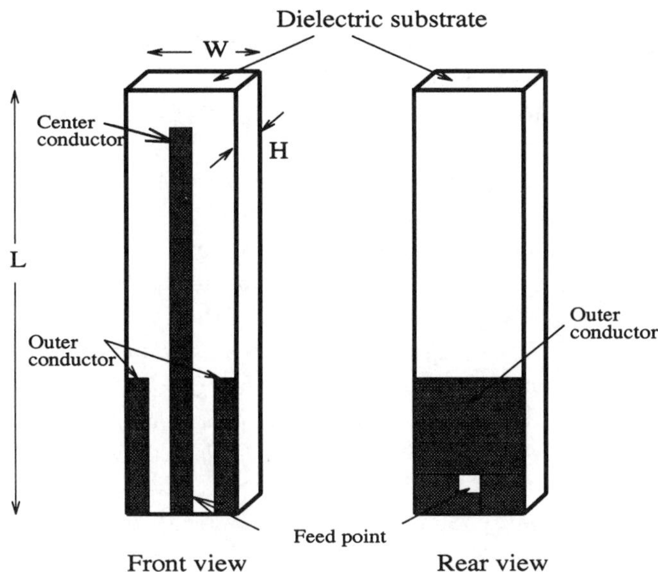

Figure 10.15 Printed sleeve antenna element.

interest. Good axial ratio is also maintained in the principal axes of the element. The drawback to this approach is that, at C-band, a small penalty in additional insertion loss may be paid. At Ku-band, the additional loss can be prohibitive [31]. This effect can be mitigated by the integration of the amplifiers directly into the radiating elements structure.

Another example of an array of microstrip patch radiators is a lightweight Ka-band 32×32 element array that uses sequentially switched scanning [32]. This highly integrated system contains three major subassemblies: a microstrip patch radiator antenna array, a transceiver, and control electronics. The microstrip elements are in a 32×32 grid within an area of 9 in^2. Polyrod directors are placed at every element. These directors enhance the gain and beamwidth of the microstrip patch antennas. The microstrip antenna element has a typical return loss of 27 dB. The entire array operates on 20W of dc power, weighs less than 5 lbs, and measures $10 \times 10 \times 5$ in.

References

[1] Y. T. Lo and S. W. Lee, eds., *Antenna Handbook, Theory, Applications and Design*, Van Nostrand Reinhold Company, New York, 1988.

[2] A. W. Rudge, K. Milne, A. D. Olver, and P. Knight, eds., *The Handbook of Antenna Design*, Peter Peregrinus, London, 1986.

[3] J. R. James and P. S.Hall, *Handbook of Microstrip Antennas*, Peter Peregrinus, London, 1989.

[4] J. D. Kraus, *Antennas*, McGraw-Hill, New York, 1988.

[5] Y. Hwang, "Satellite antennas," *Proc. IEEE*, vol. 80, pp. 183–193, Jan. 1992.

[6] A. W. Love, A. W. Rudge, and A. D. Olver, "Primary feed antennas," in A. W. Rudge, K. Milne, A. D. Olver, and P. Knight, eds., *The Handbook of Antenna Design*, Peter Peregrinus, London, 1986.

[7] G. S. Gupta, D. N. Nguyen, and C. C. Huang, "Analysis and optimization of high-performance antenna feed," *Microwave Optical Tech. Lett.*, vol. 3, pp. 286–290, Aug. 1990.

[8] E. Lier and P.-S. Kildal, "Soft and hard horn antennas," *IEEE Trans. Antennas Propagat.*, vol. 36, pp. 1152–1157, Aug. 1988.

[9] A. D. Olver, "Corrugated horns," *Electron. Comm. Eng. J.*, vol. 4, pp. 4–10, Feb. 1992.

[10] E. Lier, "A dielectric hybrid mode antenna feed: A simple alternative to the corrugated horn," *IEEE Trans. Antennas Propagat.*, vol. 34, pp. 21–29, Jan. 1986.

[11] A. D. Olver, P. J. B. Clarricoats, and K. Raghvan, "Dielectric cone loaded horn antennas," *IEE Proc., Pt. H*, vol. 135, pp. 158–162, June 1988.

[12] E. Lier and C. Stoffels, "Propagation and radiation characteristics of rectangular dielectric-loaded hybrid mode horn," *IEE Proc., Pt. H*, vol. 138, pp. 407–411, Oct. 1991.

[13] E. Lier, "Broad-band elliptical beamshape horns with low cross polarization," *IEEE Trans. Antennas Propagat.*, vol. 38, pp. 800–805, June 1990.

[14] E. Lier, S. Rengarajan, and Y. Rahmat-Samii, "Comparison between different rectangular horn elements for array antenna applications," *IEE Proc., Pt. H*, vol. 138, pp. 283–288, Aug. 1991.

[15] C. M. Knop, Y. B. Cheng, and E. L. Ostertag, "On the fields in a conical horn having an arbitrary wall impedance," *IEEE Trans. Antennas Propagat.*, vol. 34, pp. 1092–1098, Sept. 1986.

[16] A. Kumar, "Dielectric-lined waveguide feed," *IEEE Trans. Antennas Propagat.*, vol. 27, pp. 279–282, March 1979.

[17] E. Lier and T. Schaug-Pettersen, "The strip-loaded hybrid-mode feed horn," *IEEE Trans. Antennas propagat.*, vol. 35, pp. 1086–1089, Sept. 1987.

[18] E. Lier, "Analysis of soft and hard strip-loaded horns using a circular cylindrical model," *IEEE Trans. Antennas Propagat.*, vol. 38, pp. 783–793, June 1990.

[19] S. Rodrigues, P. Mohanan, and K. G. Nair, "Simulated corrugated feed horn antenna," in *IEEE AP-S Int. Symposium Digest Antennas Propagat.*, pp. 984–987, 1990.

[20] S. F. Mahmoud and M. S. Aly, "A new version of dielectric lined waveguide with low cross-polar radiation," *IEEE Trans. Antennas Propagat.*, vol. 35, pp. 210–212, Feb. 1987.

[21] R. A. Nair and R. Shafiei, "A multimode pyramidal horn with symmetrical dielectric loading—A high efficiency feed for reflector antennas in satellite communications," in *IEEE AP-S Int. Symposium Digest Antennas Propagat.*, pp. 1522–1525, 1990.

[22] R. A. Nair and K. Schroeder, "A circularly apertured dielectric loaded plural mode sectoral horn – an efficient feed for reflector antennas," in *IEEE AP-S Int. Symposium Digest Antennas Propagat.*, pp. 1534–1537, 1990.

[23] P. S. Kildal and E. Lier, "Hard horns improve cluster feeds of satellite antennas," *Electron. Lett.*, vol. 24, pp. 491–492, Apr. 1988.

[24] A. D. Graig et al., "Final report of study on digital beamforming networks," Technical Report TP8721, European Space Agency, Stevenage, England, July 1990.

[25] D. M. Pozar, "Microstrip antennas," *Proc. IEEE*, vol. 80, pp. 79–91, Jan 1992.

[26] R. J. Mailloux, "Antenna array architecture," *Proc. IEEE*, vol. 80, pp. 163–172, Jan. 1992.

[27] K. C. Gupta, R. Garg, and I. J. Bahl, *Microstrip Lines and Slotlines*, Artech House, Norwood, MA, 1979.

[28] Y. C. Shih and T. Itoh, "Transmission lines and waveguides," in Y. T. Lo and S. W. Lee, eds., *Antenna Handbook, Theory, Applications and Design*, Van Nostrand Reinhold Company, New York, 1988.

[29] N. T. Sangary, E. A. Navarro, C. Wu, and J. Litva, "Application of non-uniform FDTD method to PCS antennas," in *IEEE AP-S Int. Symposium Digest Antennas Propagat.*, June 1995.

[30] L. Sichan, R. M. Sorbello, S. Siddigi, and J. I. Upshur, "A lightweight, c-band, dual circularly polarized microstrip antenna for satellite applications," in *AP-S Int. Symposium Digest Antennas Propagat.*, pp. 909–912, 1989.

[31] A. I. Zaghloul, Y. Hwang, R. M. Sorbello, and F. T. Assal, "Advances in multibeam communications satellite antennas," *Proc. IEEE*, vol. 78, pp. 1214–1232, Jan. 1990.

[32] Northeast Microwave Systems, Inc., "A Ka-band 32x32 element scanning array," *Microwave J.*, vol. 34, pp. 149, Nov. 1991.

Chapter 11

DBF Transceiver Technology

'he performance of a DBF array is, in a large measure, determined by the capa-ilities of the transceivers that are used at the antenna elements. The transceivers erform the functions of frequency conversion, filtering, and amplification of the ignal to a power level that is commensurate with the input requirements of the .DCs or with the output power requirements. Since the elemental transceivers are significant factor in determining the cost of a DBF antenna, it is important when esigning transceivers to adopt an architecture that leads to transceivers that are w in cost and yet meet performance requirements. In this chapter, we will discuss ie architectures of transceivers, RF components, and ADC technology. We will lso include some examples of DBF systems.

1.1 ELEMENTAL TRANSCEIVER ARCHITECTURES

1.1.1 Single-Down-Conversion Receiver

single-down-conversion receiver, shown in Figure 11.1, has the simplest architec-ire. The bandpass filter is required to eliminate out-of-band interference. The andpass filter can be placed before or after the LNA with different consequences. y putting the filter before the LNA, one may obtain the highest dynamic range, hereas by placing the filter after the LNA, one can obtain the maximum receiver nsitivity.

The simplicity of this particular architecture is offset by significant deficiencies its performance. When the synchronous detector LO frequency is changed to ine the receiver over the operating band, the dc levels in the synchronous detector well as the balance between the in-phase and quadrature components may be pected to change. The output from the synchronous detector outputs may require mpensation at each operating frequency because of phase and amplitude changes at occur in the mixers due to the change in the LO frequency.

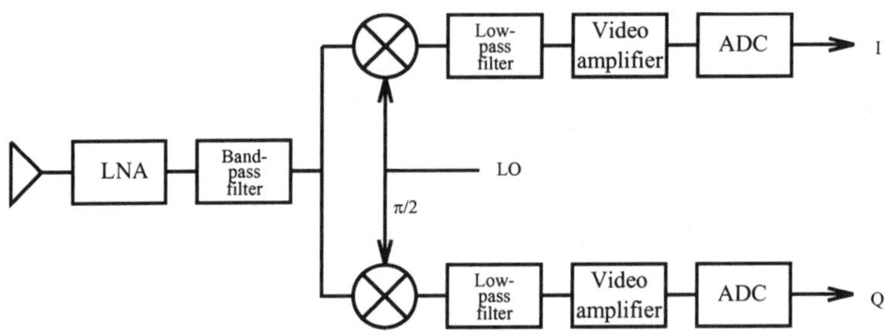

Figure 11.1 Single-down-conversion elemental receiver.

The phase and amplitude relationship between the in-phase (I) and quadrature-phase (Q) outputs of the synchronous detector is also determined by the match between the two lowpass filters following the synchronous detectors. The lowpass filters normally determine the signal bandwidth. Stringent phase- and amplitude-matching requirements across the signal bandwidth make the filters difficult to build and hence expensive.

Since only two amplifiers are used, it is difficult to realize a large dynamic range, low noise figure and stable transfer characteristics. Within the bounds of technical feasibility, it is a design requirement that the LNA and video amplifiers have sufficient gain to offset all receiver losses and raise the receiver's noise level to the quantization level of the ADCs. There can be a number of adverse factors that result from having a large gain in the LNA and video amplifiers. The dynamic range of the synchronous detectors limits the gain that can be placed in the LNA before receiver dynamic range is adversely impacted. If one opts to use high-gain video amplifiers, one runs the risk of incurring large variations in the dc imbalance of the receiver. High-gain video amplifiers usually have a poor noise figure. It is therefore undesirable to have large gain in the video amplifier. Accordingly, even though single-down-conversion receivers are attractive from a cost point of view their poor technical performance may make them unsuitable for high-performance DBF system applications.

11.1.2 Double-Down-Conversion Receiver

In a double-down-conversion receiver shown in Figure 11.2, a fixed LO is used to supply the reference signal to the synchronous detector. The synchronous detector balance and compensation are therefore independent of the agile operating frequency.

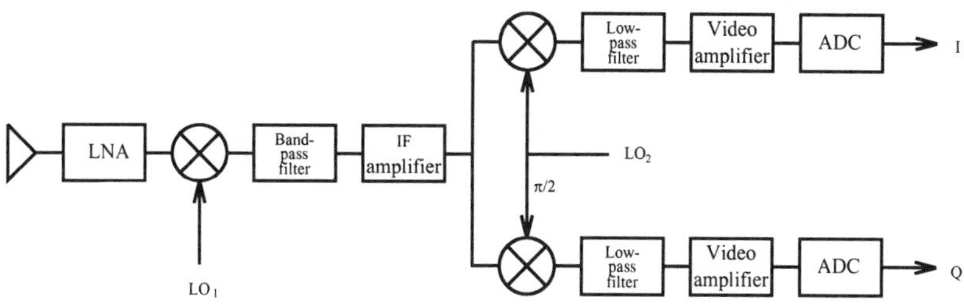

Figure 11.2 Double-down-conversion elemental receiver.

The first LO frequency is chosen such that the IF frequency of the output from he first mixer must be sufficiently high in order for the bandpass filter to reject he image frequency components resulting from the first mixer, as shown in Figure 11.3. The second LO frequency is required to be agile so that the receiver can select the desired operating frequency channel. With an additional IF amplifier n the double-down-conversion receiver, one can achieve greater flexibility in distributing gain throughout the receiver. The three amplifiers in the receiver can be configured so that the overall performance is optimized to achieve an optimal balance between high dynamic range and low noise figure. Although the match between receiver channels is improved by going to a double-down-conversion receiver, here is still significant difficulty in attaining, using hardware circuits, the channel matching needed for high-performance beamforming, such as beams with low sidelobes. However, the channel imbalance can be reduced by manipulating the digital signals using software.

11.1.3 Triple-Down-Conversion Receiver

An additional mixer can be added to the previous receiver architecture, forming a triple-down-conversion receiver, shown in Figure 11.4. This architecture allows he frequencies of the IF stages to be optimized for specific functions. The second IF frequency can be made sufficiently low so that a bandpass filter, such as a surface acoustic wave (SAW) filter, can be used to define the signal bandwidth. Since a SAW filter is a finite impulse response (FIR) filter, it has very low phase and amplitude distortion. SAW filters can be built to be closely matched, yielding elemental receiver responses that are highly correlated. The lower operating frequency for the synchronous detector makes the detector less sensitive to variations in the LO distribution. The minimum operating frequency of the synchronous detector is defined by the signal bandwidth and the performance of the low-pass filters in

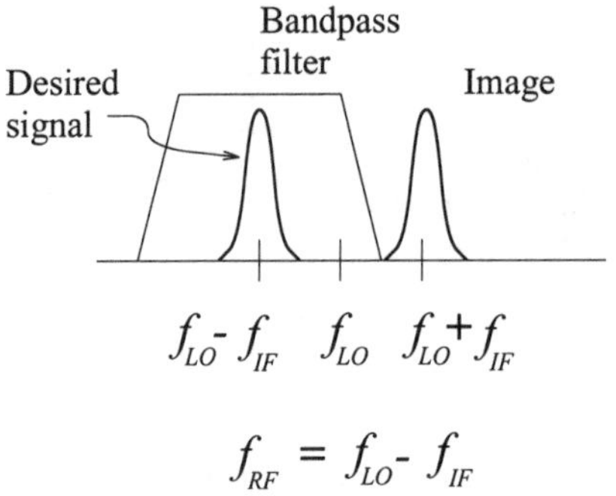

Figure 11.3 RF signal and its image frequency component at the output of the mixer.

suppressing the leakage between the LO and the RF inputs and the dc output of the synchronous detector mixers. They are made much wider than the signal bandpass to enssure that they have virtually no impact on receiver matching.

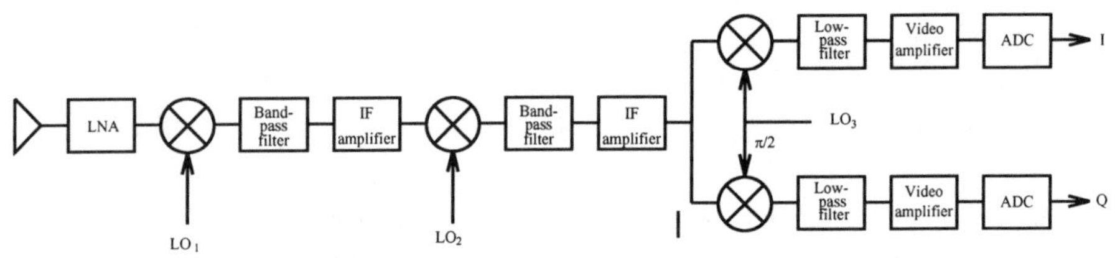

Figure 11.4 Triple-down-conversion elemental receiver.

Since the filter bandwidth and center frequencies of the IF stages are all interrelated, compromises may be required in the choice of the parameters. The bandwidth of the first IF stage is a major factor in determining the parameters of the next IF stage. The IF frequencies should be selected to meet the image rejection requirements for the filters. If one starts with a high frequency for the first IF stage, it is relatively easy to design the first bandpass filter to meet the image rejection requirements, but it makes it relatively difficult to design the bandpass filter in the second IF stage to meet the image rejection requirements. The bandwidth of the second bandpass filter is required to be narrow enough to suppress noise appearing in the image response of the second IF mixer. Therefore, the frequency for the second IF stage should be high enough for the bandpass filter to be realized in practice.

The triple-down-conversion receiver is flexible to be configured to have large dynamic range, low noise figure, and relatively stable transfer characteristics. It would therefore appear to be the preferred receiver structure for DBF communications systems. For space applications, however, the triple-down-conversion receiver may present a number of problems. Due to the need for additional components, the weight of the payload may increase, and because of the lower efficiency of the receivers, power consumption may also increase.

11.1.4 Direct Sampling Receiver

The three receivers discussed previously have a number of deficiencies in common. The shortcomings all stem from a common root; that is, the fact that the baseband I and Q signals are all derived from mixing the last IF signal with a reference LO. This means that the overall bandwidth in the receiver is defined by either the last IF bandpass filter or alternatively by the lowpass filters in the I/Q base band signals. Because of the way in which the I and Q signals are derived, they may suffer from the following performance limitations:

1. There may be a poor match between the characteristics of the I and Q signals over the received bandwidth.
2. The I and Q channels may not maintain phase quadrature over the receiver bandwidth.
3. The I and Q channels may have separate dc offsets.
4. Nonlinearities of the components used in the I and Q channels may produce spurious noise.

The amplitude matching and the phase orthogonality of the I and Q channels are extremely critical and tend to be a major source of error. This has created interest in an alternative conversion technique (Figure 11.5), in which the signal is directly sampled and digitized at IF and the complex video signal is generated digitally. This approach eliminates the analog lowpass filters, analog video amplifiers, and two ADCs in the previous receivers at the expense of one faster ADC and some additional digital circuitry.

Since the I and Q outputs are formed using a digital filter in the direct IF conversion receiver, I and Q matching problems are effectively eliminated from the channel.

There are two methods for carrying out the direct sampling: Nyquist sampling and down-sampling (also called direct down-conversion). The general Nyquist sampling theorem for sampling a bandpass analog signal (a signal having no frequency components above a certain frequency $f_{\max}$) requires that the sampling rate be at least two times the highest frequency component of the analog signal (i.e., $2f_{\max}$).

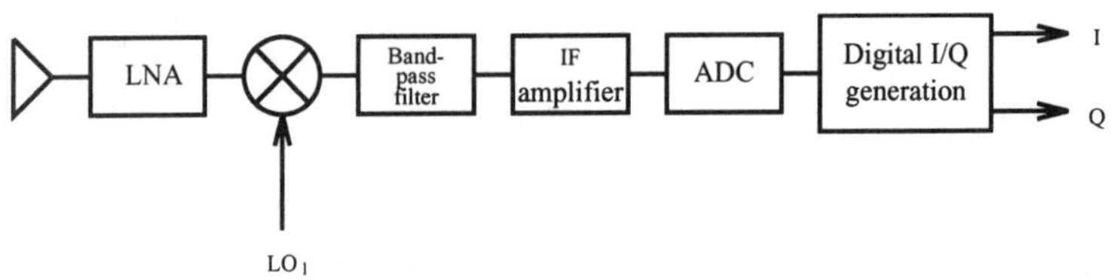

Figure 11.5 Direct sampling receiver.

For an IF signal with a bandwidth B centered at $f_{\rm IF}$, the sampling rate mus then be $2(f_{\rm IF} + \frac{B}{2})$. A digital down-converter (DDC) is required to carry out th coherent phase detection necessary for generating the digital I and Q signals. A DDC contains a synthesizer, a quadrature pair of digital multipliers (which act a mixers in the digital domain), and some clever filters that implement both low-pas and decimation. The device extracts a narrowband signal from a wideband digita input and decimates it to a reduced data rate. Since the new data rate is now proportional to the bandwidth, all of the information in the digital signal has bee captured without superfluous bandwidth being carried by the digital signal. Th first benefit derived from the reduced data rate is that the digital signal processo is better able to cope with it. Figure 11.6 shows a block diagram of a DDC. A a matter of fact, in this approach, the down-conversion process is carried out i digital domain instead of the analog one. Of course, this approach requires ver fast ADCs and digital multipliers.

It is possible to use sampling at rates lower than $2f_{\rm max}$ and still get an exac reconstruction of the information contained in the analog signal in the bandpas signal case. This approach is referred to as direct down-conversion. For a bandpas signal, the minimum sampling rate that allows for extraction of the informatio carried by the signal is two times the bandwidth B. In order to ensure that spectra overlap does not occur, the sampling frequency f_s must satisfy [1]

$$\frac{2(f_{\rm IF} + B/2)}{k} \le f_s \le \frac{2(f_{\rm IF} - B/2)}{k-1} \qquad (11.1$$

where k is restricted to integer values that satisfy

$$2 \le k \le \frac{f_{\rm IF} + B/2}{B} \qquad (11.2$$

and $B \le F_{\rm IF} - B/2$. Bandpass sampling holds promise for DBF application when the desired input signals are bandpass signals. It allows ADCs with slowe sampling rates to be used in applications where performance, power consumption and cost are critical. An important practical limitation, however, is that the ADC must still be able to effectively operate on the highest frequency component in the signal. This specification is usually given as the analog input bandwidth for the ADC. Conventional ADCs are designed to operate on signals having maximum frequencies of one-half the sampling rate. Performance of the ADC typically degrades with increasing input frequency. When ADCs are used in bandpass sampling applications, the specifications of the converter must be examined to determine the behavior at higher frequency input. In addition, with bandpass sampling, stringent requirements for analog bandpass filters (e.g., steep rolloffs) are needed to prevent distortion of the desired signal from strong, unwanted adjacent frequency components. Furthermore, the sampling jitter requirements of the sampler are high to ensure that the carrier frequency is accurately sampled to recover the signal phase information.

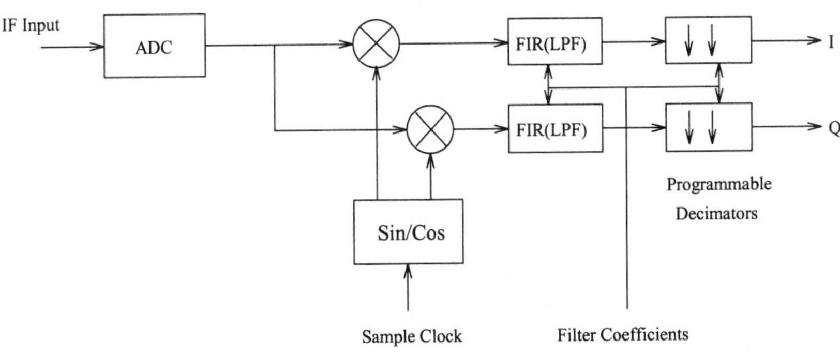

Figure 11.6 Digital coherent phase detection to generate I/Q signals.

11.1.5 Software Receiver

The advantages of the direct sampling receiver stem from the fact that digital coherent phase detection is used to generate I/Q signals. In fact, this approach follows the philosophy that ADCs should be used as close to the antenna as possible, with as much radio functionality as possible defined in software. The natural step forward from direct sampling will be placing the ADC right after the RF frontend, as shown in Figure 11.7. The other radio functions such as IF, baseband, and bit stream are carried out using digital programmable processors. The resulting software-defined receiver, which is the key component of a *software radio* system [2], extends the evolution of programmable hardware, increasing flexibility via the total programmability of RF bands, channel access modes, and channel modulation. The underlying flexibility allows for multiband and multimode operation, which is very necessary for the seamless integration of PCS, land mobile, and satellite mobile services. Furthermore, for the same reason, it also allows the maximum possible DBF performance.

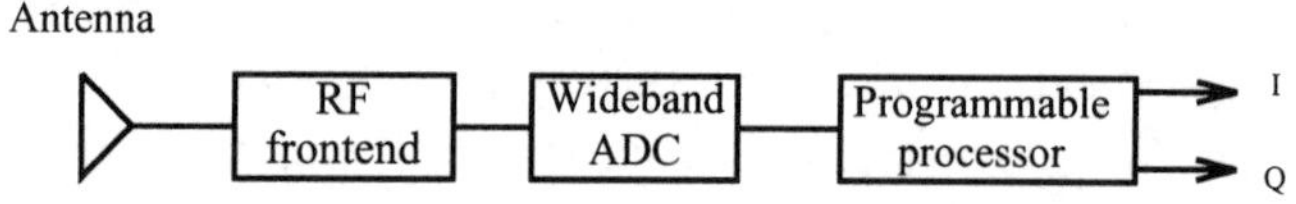

Figure 11.7 Software-defined receiver.

There are, however, a number of technological challenges that must be addressed. The critical one is the availability of affordable wideband ADCs, which will be discussed in the next section. In addition, it is difficult to accurately estimate the processing demand of applications and processing capacity of programmable DSP configurations. That is, an open architecture standard for high-capacity wideband signal buses has yet to emerge. High-performance digital signal processors are also essential in the development of practical software receivers.

11.1.6 Transmitter

The function of a transmitter is to amplify an RF carrier modulated with the desired signal, adding minimal distortion to the encoded information. A typical transmitter functional block diagram is shown in Figure 11.8. The lowpass filter is used to eliminate harmonics of the transmitted signal. Since dynamic range and sensitivity are not considerations in transmitter design, usually one stage of up-conversion is required; that is, the I and Q baseband signals are directly mixed with the RF LO carrier. Instead, the considerations in transmitter design are a different set of parameters:

1. *Power output* is the basic requirement of a communications system and is defined as the rms power in a PM or FM system and as the peak envelope power in a AM system. The power output depends largely on the characteristics of the power amplifier.
2. *Spurious outputs* consist mostly of harmonics of the carrier. Spurious outputs are usually 70 to 90 dB below the carrier.
3. *SNR of the transmitted signal* is the quality measure of the transmitted signal. Noise components usually result from thermal noise in the transmitter circuits, power supply hum, noise pickup, and nonlinearities in the transmitter circuitry.

4. *Transmitter turn-on time* is defined as the time for power output to reach 90% of its specified power level. For digital communications systems, especially TDD systems, transmitter turn-on time is an important factor. It must be sufficiently short to meet the throughput requirement, but not so short to cause transient interference to adjacent channels.
5. *Adjacent channel interference* is caused by a number of sources, such as the single-sideband (SSB) phase noise of the LO, the modulation process, and a short transmitter turn-on time.

In addition to the above considerations, the balance between transmitter channels must be taken into account in DBF applications.

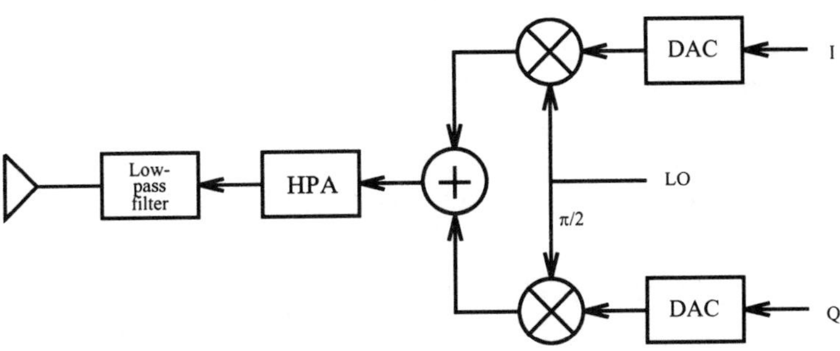

Figure 11.8 A typical transmitter.

11.1.7 Internal Calibration Loop

Internal calibration loops are used to measure the imbalance between the I and Q components of a transceiver and the imbalance between transceiver channels in order to eliminate these imbalances using digital signal processing. To measure the imbalances in the receiving channels, identical test signals are injected via the couplers to the individual receivers periodically, as shown in Figure 11.9. The measured values of the I and Q components are compared with the desired values and their discrepancies will be compensated for by adjusting the receiving weights in the beamforming process. The SNR of the test signals at the outputs of the receivers must be high enough to allow an accurate estimation of the imbalances. To measure the imbalances in the transmitter channels, I and Q test signals are input to the transmitters and the transmitted signal will be captured by the couplers and fed via the bypass circuits to the receivers. This requires, of course, that the receivers be calibrated first. The measured values of the I and Q components will be analyzed to determine the discrepancies with respect to the desired values. The signal processor will adjust the transmission weights that minimize the discrepancies.

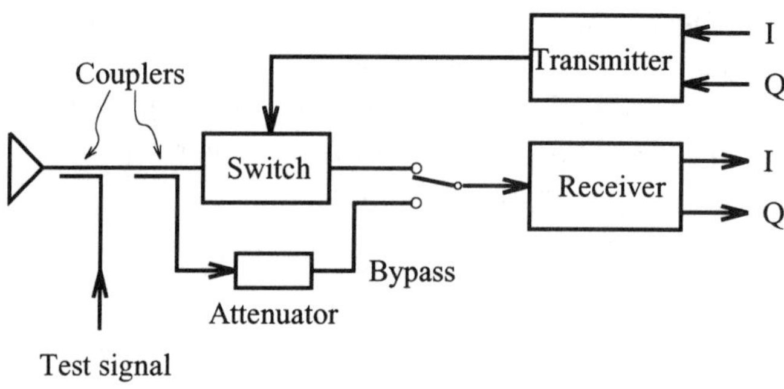

Figure 11.9 Internal calibration circuit.

11.2 RF COMPONENTS AND MODULES

The RF components play a significant role in the performance of the complete DBF antenna array. In this section, we present the state of the art of these RF components in terms of their characteristics and performance, which may be helpful in designing a DBF system. However, due to the limited scope of this text, we do not intend to discuss these components at the circuit level.

11.2.1 Power Amplifiers

An MMIC power amplifier makes use of integrated metal semiconductor field-effect transistors (MESFET) in the output stage, which are configured to handle large voltage and current swings. The gate length and layer parameters needed to achieve large-signal capability are normally very different from those required for achieving low noise and high gain.

Of all the different types of monolithic components, the power amplifier is one of the most important, but also most difficult to fabricate. The difficulties arise because of the high-thermal impedance of the semi-insulating substrate used in their devices. Therefore, the design challenge is to develop a suitable thermal management scheme.

In order to increase the output power, one may consider using a large-gate periphery device. This in turn may result in low input and output impedance. The transforming networks that are then required to achieve 50 Ω matches may be implemented on-chip or off-chip [3].

Broadband MMIC power amplifiers are customarily designed using one of the following three formats [4]:

1. Single-ended units that are applied as balanced stages using Lange couplers to provide acceptable levels of I/O impedance match;
2. Distributed amplifiers that can be applied as single-ended gain blocks;
3. Feedback amplifiers that can also be used in single-ended configuration.

It is currently common practice to use single-ended units as power-driver-stage devices, distributed amplifiers as either power-driver or low-noise devices, and feedback amplifiers as small-signal units. Single-ended designs provide a bandwidth of 1 to 1.5 octaves, while distributed and feedback designs have demonstrated an instantaneous bandwidth of up to a decade in frequency. Table 11.1 compares some of the features of single-ended and distributed MMIC power amplifiers. When one is deciding on the design of devices to be used for a particular application, it is important to consider the cost and gain advantages that result from the use of Lange couplers. These can be used to turn single-ended stages into balanced stages, with all of the attending advantages.

Table 11.1

Comparison of Single-Ended and Distributed Amplifiers

	Single-Ended	*Distributed*
Chip size	small	large
Efficiency	moderate	low
Matching circuits	complex	simple
I/O standing wave ratio (SWR)	typically 2	1.5 (max)
Configuration	balanced	single-ended

GaAs MMIC power amplifiers are currently being considered for systems such as DECT and PHP. A few examples of power amplifiers are given in Table 11.2.

11.2.2 Low Noise Amplifiers

LNAs also benefit from monolithic integration. Excellent results have been reported. Because LNAs are used as receiver preamplifiers, low-noise performance is the first requirement. In addition, an LNA is also required to exhibit a flat, well-controlled gain response, good phase linearity, and good terminal voltage standing wave ratios (VSWR) over a broad bandwidth.

Table 11.2
Examples of GaAs MMIC Power Amplifiers

Maker	*Output Power*	*Frequency*	*Application*
M/A-COM	1.4W	14 to 14.5 GHz	VSAT
Celeritek	33 dBm	1.95 GHz	PCS
Pacific Monolithic	30 dBm	1.9 GHz	PCS

Three generic types of technologies can be considered to fulfill these require-ments, namely, reactively matched [5], shunt feedback [6], or traveling-wave [7 amplifier circuits. When reactively matched circuit elements are used the design o a system, it is usually found that system performance is sensitive to the values of th passive components and variations in FET parameters. Furthermore, the matchin networks used at the input and output of multisection stages, as well as those use between stages, make gain flatness very difficult to achieve and also consume larg area of GaAs. In contrast, a design based on shunt feedback often provides opti-mum performance trade-offs for small- and medium-signal level amplifiers of 1 to octave bandwidths. The merits of using feedback in circuit design are excellent gai flatness, amplifier stability, and good matches at input and output, all within a rel-atively small areas of GaAs. Considering the advantages of attractive technologies the initial choice should be weighted in favor of shunt feedback topology because i offers the best solution for meet the requirements for high-quality LNAs.

The characteristics of the FET gate periphery used the design of the LNAs i quite important in determining their ultimate performance. Several factors must b taken into consideration, including noise figure and associated gain, circuit stability dc power consumption, input and output impedance, sensitivity to mismatches, an output power capabilities. A few examples of LNAs are given in Table 11.3.

11.2.3 Mixers

The development of broadband MMIC-based mixers represents a significant chal-lenge. Single-ended as well as balanced and double balanced mixers can be designe using dual-gate FETs. Dual-gate devices offer several advantages over conventiona devices such as ease of LO injection, improved isolation, and added gain.

Typical performances of mixer chips reported recently are as follows:

1. Coplanar waveguides, slotlines, and coplanar strips are combined to realiz an MMIC double-double balanced mixer (DDBM) [8]. The DDBM exhibit

Table 11.3
Examples of Low-Noise Amplifiers

Maker	*Gain*	*Frequency*	*Noise Figure*
M/A-COM	20 dB	1.7 to 2 GHz	$\tilde{2}$ dB
AML Communications	-6 to 16 dB	824 to 851 MHz	<3.0 dB
TriQuint Semiconductor	25 dB	500 to 2500 MHz	1.3 dB
MITEQ	40 dB	1 to 2 GHz	1.4 to 2 dB

RF, LO and an IF bandwidth of 6 to 20 GHz, 8 to 18 GHz and 2 to 7 GHz, respectively, with conversion loss ranging between 6.2 and 9.8 dB, and RF to IF, LO, to IF and LO to RF isolation all greater than 20 dB.

2. An integrated W-band down-converter was fabricated using a 0.15 μm T-gate lattice-matched InGaAs/InP high-electron-mobility transistor (HEMT) [9]. The mixer showed a 2.4-dB conversion gain and 7.3-dB noise figure at 94 GHz. The complete down-converter exhibited a 3.6-dB noise figure and 17.8-dB conversion gain at the waveguide input.
3. Recently, a Si down-converter IC was developed. Using a novel 0.6-μm Si bipolar process, called DNP-III for an f_t of 20 GHz, more than 10 dB of gain was claimed in the 1- to 2-GHz frequency range [10]. Typically, power dissipation was measured to be 125 mW.
4. A low noise block down-converter chip designed for direct broadcast satellite (DBS) receivers was reported [11], which operated from 11.7 to 12.2 GHz. It costs less than 10 dollars. The chip is housed in a hermetic package and must be used in conjunction with a dielectric resonator to be fully functional.

11.2.4 Oscillators

Microwave oscillator design is similar to microwave amplifier design. The oscillator design may have the same dc-biasing circuits and the same active device with the same set of S parameters used for amplifier design. For example, negative resistance is commonly used for amplifier design, and it is also used for oscillator design [12]. There are also other techniques in MMIC oscillator design, such as single-ended [13] and push-pull [14] techniques. It can be found that a buffer amplifier at the oscillator output can usually provide better load isolation and power output stability.

It is well known that MMIC-based oscillators have very high noise power density and that the stability is rather poor [15]. A common solution to this problem is to utilize a resonator external to the MMIC. There are a number of choices for the external resonator, such as a dielectric resonator oscillator (DRO) [16], a crystal yttrium iron garnet (YIG) oscillator, or a phase-locked loop (PLL) with an external oscillator [15].

Typical performance characteristics of recently developed oscillator chips are as follows:

1. A Ku-band MMIC voltage-controlled oscillator (VCO) was reported in [15]. The oscillator is tunable over the frequency range 14 to 18 GHz with a minimum output power of 18 dBm and a phase noise of -88 dBc/Hz at a 100-kHz offset from the carrier over the temperature range 0 to 65°C.
2. A wideband MMIC oscillator chip was described by [16], which was designed for fast-switching DRO and VCO applications. Its frequency settles within 0.6 MHz of the final frequency in only 0.5 μs. As a VCO, the frequency range of this device varies from 10 to 14 GHz.

11.2.5 Switches

Broadband switches represent serious problems for the designer because of the wideband requirements for these devices. The designers attempt to incorporate as many devices as possible with characteristics that are highly insensitive to frequency changes. We demonstrate that acceptable solutions are available by describing some representative devices and listing their performance.

The broadband switches that are used in switching circuits all currently use FETs employed in a "resistive" switch mode. The virtues of the concept are that the switch consumes negligible switching or "hold" power, and switching speeds are typically less than 1ns. In addition, the FET switches are entirely compatible with the fabrication processes used to form amplifiers and are consequently highly integrable. On the negative side, when compared to PIN switches, the FETs provide relatively high insertion loss. When they are used as input or lower power T/R switches, the insertion loss penalty is not important unless a premium is being placed on noise figure. When they are used as output switches, where switching operations have been demonstrated with power levels of a few watts, the typical insertion loss encountered, 0.7 to 0.8 dB, imposes a heavy tax on delivered power.

Table 11.4 gives the performance data for a number of GaAs MMIC switches, which would be used in broadband system applications [17–21].

Table 11.4
Typical Performance of MMIC Switches

Freq. Range (GHz)	*Isolation (dB)*	*Insertion Loss (dB)*	*Input VSWR*
2–18	80 (min)	2.3 (max)	1.6:1 (max)
dc–20	35 (min)	1.6 (max)	1.5:1 (max)
30–40	32 (min)	0.7 (max)	1.5:1 (max)

11.2.6 T/R Modules

Multichip T/R Modules

The advantages of GaAs MMICs over the traditional hybrid-circuits approach are well established. However, two obstacles stand in the way of widespread use of MMICs in wireless systems. They are high MMIC development costs and the long development cycle. GaAs macrocells offer a quick, low-cost and low-risk approach to new MMIC system design. The crux of the macrocell-based design methodology is the use of precharacterized component cells (e.g., amplifiers, mixers, switches, phase shifters) to form system-functional building blocks. These building blocks are integrated into larger GaAs ICs to form a single-chip subsystem, which can, in turn, be combined to build complete microwave systems [22].

A typical block diagram of the T/R module is given in Figure 11.10. To realize a complete MMIC T/R module, certain critical problems must first be solved. One is the elimination of the requirement for narrowband filters in the system design. An MMIC filter occupies a large area and limits the usable frequency range. The use of a filter fails to exploit the MMIC's distinctive features. Since the RF amplifier has a broadband response, undesired signals or noise signals corresponding to the image frequency may inevitably be amplified and converted along with the wanted signal. To reduce the undesired image noise, without resorting to filters, one will find that an image rejection mixer is indispensable.

Another problem faced by the design of a MMIC T/R module is the suppression of MMIC oscillator noise. An MMIC oscillator has a fairly noisy output spectrum and its monolithic resonator has low-Q characteristics. Oscillator noise consists of phase noise near the carrier, as well as noise removal from the carrier frequency. The phase noise problem is solved by employing a high-speed PLL. Since phase noise is the result of rapid fluctuations in the oscillator frequency, it can be significantly suppressed by expanding the loop bandwidth of the PLL circuit. The following are some examples of typical multichip MMIC transmitter or receiver modules and the

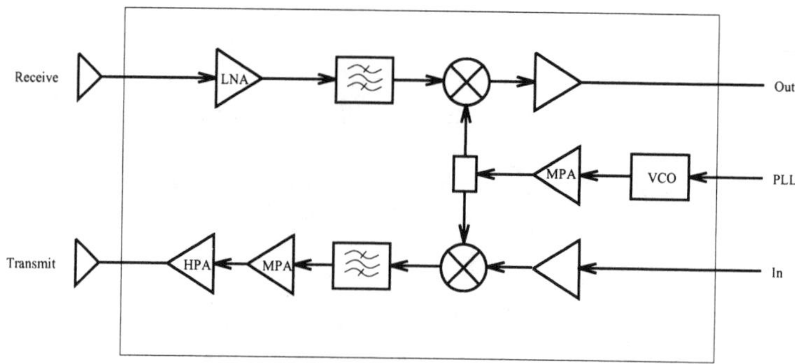

Figure 11.10 Configuration of an MMIC T/R module.

values of their operational parameters:

1. A 30-GHz MMIC for satellite transponder receiver was implemented using seven MMIC modules plus a crystal oscillator and a loop circuit (NTT, Japan, 1990, [23]). Some of its operational parameters are listed in Table 11.5.

Table 11.5
Operation Parameters of a 30-GHz MMIC Receiver by NTT [23]

	Values	*Units*
Receiving freq.	29.25 to 31.0	GHz
Small signal gain	98 (min)	dB
Noise figure	8.2 (max)	dB
Limiting range	27 (min)	dB
IF	1.045 ±0.1	GHz
IF output level	10 ±0.7	dBm
Image ratio	18 (min)	dB
AM/PM conversion	1 (max)	deg /dB
Local oscillator	-80 (max)	dBc/Hz
Spurious responses	50 (min)	dB
Power consumption	5.2	W
Operation temperature	-10 to 40	°C

2. An S-band receiver frontend incorporates five MMIC chips, including low-noise amplifier, in-phase signal splitter, quadrature local oscillator splitter, passive routing chip and dual-channel mixer (Plessey, U.S., 1985, [24]). The receiver system has a gain of 36 dB and operates in the band 2.7 to 3.1 GHz.

The maximum noise figure is 8 dB and the IF output level at 60 MHz is 11 dBm.

3. An L-band MMIC receiver was developed by ANADIGICS, Inc. [25]. The front end of the receiver was comprosed of a single-pole double-throw switch, which was used for antenna selection, followed by RF amplification, double balanced mixer, and buffer amplifier (for the local oscillator). The first IF amplifier, which is preceded by a SAW filter, is followed by another filter and a second high-gain limiting amplifier. The receiver operates between 0.8 to 1.2 GHz and has a gain of 100 dB. The IF is at 300 MHz.
4. A receiver system composed of two LNAs, a Schottky diode mixer, and a buffer amplifier, which was fabricated using 0.25-μm PMHFET technology has been integrated in a package with K connectors [26]. A conversion gain of 24 dB along with a double-side-band noise figure of under 4 dB has been measured at 38 GHz. The system has a dynamic range of 25 dB and has an output level of 20 dBm at the 1-dB compression.
5. An X-band T/R module for space applications has been designed and fabricated by Alenia Research Center [27]. This module, which was developed based on the multichip module approach, was integrated in a single package. the package incorporates thin-film interconnections and bias, as well as driver and switching circuits. A relatively small number of MMIC was used in the module. The RF and dc interconnections along with the Si integrated circuits for signal control was realized using a highly integrated form of thin-film deposition on aluminum. The whole ceramic package is hermetically sealed and it has feedthrough ports for RF, dc and control signals.

Fully Monolithic T/R Modules

Fully integrated MMIC T/R modules improve system reliability and reduce assembly and tuning time because of the reduction in parts count and interconnections. Unit-to-unit phase and gain tracking is also improved for the same reasons. These attributes make MMICs an excellent choice for applications where repeatable and reliable performance is required. The fully integrated MMIC system chip is, no doubt, the best choice of the available options for wireless communications, especially for satellite communications. There are a number of fully integrated MMIC T/R modules, that have been reported in the literature, such as:

1. A fully integrated GaAs microwave receiver was developed for DBS applications at 12 GHz. The receiver comprised an LNA, filter, mixer, LO, and IF amplifier on a single GaAs chip. The LO was a common-source FET oscillator, configured with series capacitive feedback circuits, which were used for

stabilization. It also incorporated an off-chip dielectric resonator. The technology wa based on self-aligned FETs with a 0.7-μm gate length [28]. Its operational parameters are listed in Table 11.6.

Table 11.6

Operational Parameters of a GaAs Fully Integrated Microwave Receiver

	Values	*Units*
RF	11.7 to 12.5	GHz
IF	0.95 to 1.75	GHz
Reject image band	9.0-9.8	GHz
Gain	25±3	dB
NF	4.5	dB
Chip size	2.5×2.5	mm^2

2. Finally, we describe a single chip Ka-band transceiver for FM-continuous wave (CW) radar applications. The transceiver consists of seven active microcells: VCO, buffer amplifier, power amplifier, LNA, single balanced mixer, voltage-controlled phase shifter, and attenuator. The technology is based on single-process InGaAs HEMT [29]. Its operational parameters are listed in Table 11.7.

Table 11.7

Operational Parameters of a Single-Chip Ka-Band Transceiver

	Values	*Units*
RF	37 to 40	GHz
IF	10 to 100	MHz
Conversion loss	0	dB
NF	9	dB
Transmitted output power	+12	dBm
Chip size	4.35*7.0	mm^2

11.3 ANALOG-TO-DIGITAL CONVERTERS

As mentioned earlier, the ADC is a major component of a digital beamforming system, which makes the attending signal-processing operations possible. At this junction, a review of the state of the art in ADC technology and projected future development are appropriate to the discussion of DBF. Although essentially there are only two kinds of ADCs (i.e., linear and nonlinear), the focus here will be on linear ADCs. A variety of architectures are available for today's ADCs. We will avoid going deeply into the architectures of various types of ADCs in order to give a clear picture of the state of the art in ADC technology and its possible future development.

Many parameters can be used to characterize the performance of an ADC; these are resolution, speed, linearity, dynamic range, and power consumption, just to name just a few. Usually, different applications will place different weights on the relative importance of the ADC parameters. For example, linearity may be the key specification for ADCs in a wide-bandwidth spectral signal analysis, whereas for an airborne radar application, a high dynamic range is more important [30]. As a matter of fact, as ADCs become more and more application-specific, many more new parameters are introduced to characterize ADC performance, and many of these parameters are often interrelated. After all, a designer's first consideration for an ADC is to ensure that the converter can provide data with the required resolution and throughput. An ADC's throughput is defined here as the maximum number of conversions per second that the ADC can perform. Analog-to-digital conversion used to be carried out by a sample and hold amplifier (SHA) followed by an ADC. Nowadays, ADCs usually have their own internal SHA. This makes the SHA's sampling rate an indicator of a ADC's throughput ability. In what follows, we will use sampling rates and resolution as the two prime parameters in assessing various ADCs. Of course, an ADC is used in a data acquisition system, other parameters concerning a ADC's dc and dynamic performance should always be considered along with these two prime parameters.

11.3.1 The State of the Art

The sampling rates and resolutions provided by present-day ADCs cover a wide range of values, with typical resolution values varying from 8 to 24 bits, and sampling rates ranging from a few kHz to several GHz. Invariably, ADC designers must aim for a balance between, these two parameters because it is very difficult to achieve both a high sampling rate and a high resolution.

Table 11.8 lists some representative examples of high-performance ADCs, each with its own resolution and sampling rate, that represent the state-of-the-art ADC

Table 11.8

Performance of State-of-the Art ADCs Reported in Literature.

Designer	*Resolution* (bits)	*Sampling Rate*	*Fabrication Technolo*
HP Co.	8	4 GHz	Hybrid circuits
T. Ducourant et al.	4	3 GHz	GaAs
F. G. Weiss	8	1 GHz	GaAs
Philips Research Labs	8	650 MHz	Bipolar
Kimura, Hiroshi	10	300 MHz	Bipolar
Signal Processing Tech	12	30 MHz	Bipolar
Karanicolas et al.	15	1 MHz	BiCMOS
Y. G. Ming et al.	16	320 kHz	
Crystal Semiconductor	20	256 kHz	3μm CMOS

technology [31–38]. It should be pointed out that of the ADCs listed in the table, only the one from Signal Processing Technologies Co. is commercially available. The reason is that there is a big gap now between the technology-based ADC research being carried out at various research laboratories in the world and those technologies being adopted by semiconductor companies. As Frank Goodenough pointed out in [39], most commercially available ADCs are now the product of market demand rather than technology push.

Despite this diversity of ADCs, they can be divided into two main categories: silicon-based and GaAs-base, with the silicon-based being dominant. Silicon-based ADCs usually incorporate bipolar technology, complementary metal oxide semiconductor (CMOS) technology, or the combination of the two.

The dominance of silicon-based ADCs is based on economical nature rather than technological considerations. Experiments that have been carried out on both silicon and GaAs devices have demonstrated that GaAs devices have the following advantages over silicon devices:

1. GaAs devices switch at faster rates; therefore, they pass signals more rapidly.
2. GaAs devices consume less power at high speeds.
3. GaAs devices operate over a wider temperature range.
4. GaAs devices are intrinsically more radiation-resistant (hard) than silicon devices, because greater energy is required to ionize the material.
5. GaAs devices have lower noise thresholds.

Although the unusual properties of GaAs have been recognized by users of high-speed digital and analog devices for many years, its development was slow due to the difficulties and the high costs involved in the fabrication of GaAs devices. GaAs hybrid microwave integrated circuits (HMIC) and GaAs MMICs are the only exception to this rule because of their strategic importance to the defense industry.

11.3.2 Future Trends

Predicting the trend for a rapidly moving field such as microelectronics is difficult. The history of the development of microelectronics has shown that precise long-term prediction is virtually impossible.

Although many high-speed heterostructure devices, called second-wave devices such as quantum electron devices, InP-based transistors, and high-frequency resonant-tunneling devices, have been under investigation, there is little possibility that these devices could be incorporated into the design of ADCs in the near future [40]. It is expected that silicon and GaAs will still be the two dominant technologies for ADCs.

While both silicon and GaAs technologies will continue to evolve, GaAs is expected to play an increasingly dominant role in the ADC industry. Today's high-speed ADCs still have a number of unsolved problems, primarily with respect to power consumption, dynamic range, and cost, This is the main reason why much of the current research in ADC technology is largely focused on overcoming these shortcomings. It is expected that future ADC development will be channeled into two directions: (1) improvements in speed and resolution, and (2) improvements in other areas, such as power consumption, to make ADCs easier to use.

In general, the current GaAs technology is used to extend to capabilities of low-speed silicon based systems, thereby allowing for interfaces to be made to GHz logic and microwave systems [41]. This indeed applies directly to ADCs that are required for directly sampling RF signals in a software receiver. As well, because of the rapid progress being made in fields such as signal processing, fiber optics, satellite communications, and radar, ADCs are required to perform at a much higher speed and resolution while at the same time consuming less power. In applications requiring high-speed and low power consumption, GaAs may eventually replace silicon. It may also be able to perform the functions that are currently not available in silicon. For low-speed applications, on the other hand, cost considerations may continue to tilt the balance in the direction of silicon.

It is almost certain that in a few years there will be a wide array of devices, centered on low power consumption (low supply voltage), high resolution, satisfactory dynamic range available to the design engineers. The sampling rate for these devices will be as high as 4-GHz. As for the range above 4 GHz, we could anticipate practical ADCs making their appearance in the commercial marketplace in due course. Eventually, these ADCs will be priced low enough for them to be adopted into practical systems.

11.3.3 ADC Technology and Digital Beamforming

Here we consider the ADC technology that is most likely to be used for digital beamforming. Base on the estimate made above and considering the unique requirements of digital beamforming, we can easily say that GaAs ADCs are the most promising. This is based on the following considerations [39, 42, 43]:

1. GaAs ADCs can deliver the high speeds required by digital beamforming with low power consumption. Although it is likely that some silicon-based ADCs may satisfy the speed requirements, power consumption will continue to be a severe drawback for silicon-based ADCs. This is due to the higher intrinsic power consumption of silicon devices. For example, in the 8-bit, 4-GHz silicon-based ADC listed in Table 11.8, which consists of one sampler chip, four ADC chips, and four FISO chips, the power consumption of the sampler chip alone is 5.5W. Although we can expect the power consumption of silicon-based ADCs to decrease with time, whether or not they will attain rates as high as a few GHz while at the same time drawing low levels of power is very much in doubt.
2. GaAs ADCs can operate in more adverse environmental conditions than silicon-based ADCs. They can withstand greater temperature ranges, they exhibit greater radiation hardness, and they have small noise figure. Radiation hardness or radiation resistance is a very important consideration in the design of satellite-based systems. Therefore, in the application of digital beamforming to future satellites it would be expected that GaAs will be the technology of choice for ADCs.
3. Since GaAs technology has now reached a reasonable level of maturity, it is expected that there will be steady growth in the number and selection of MMICs that will available to design engineers. Although it is still technically impossible to integrate an ADC and an MMIC onto a single monolithic IC, because the fabrication processes for the two devices are somewhat different, it is expected that with future developments in GaAs fabrication technologies, the two will eventually be integrated. The advantages of a monolithic ADC and MMIC chips are obvious:

    a. Improved circuit reliability, brought about by a reduction in the number of wires and interconnections;

    b. Promise of cost reductions due to batch processing;

c. Smaller package sizes, resulting from the use of monolithic IC technologies.

11.4 SOME EXAMPLES OF DBF SYSTEMS

A sampled aperture radar system with 32 10-dB standard gain horns at X band, a double-conversion transceiver, and analog I and Q channels was reported in [44]. The receivers are coherent, with frequency stability to 10^{-12}, accuracy of 0.1 dB and 1°, 0.1-Hz Doppler resolution, and 1-Hz to 2-kHz sampling rate. The transmitter radiates at two frequencies simultaneously, using either horizontal or vertical polarization. The power of the transmitter is 100 mW. The radiating face of the antenna consists of a linear array, composed of 32 elements (22-dB horns) and an aperture of 1.82m. An interelement spacing of 0.05715m was used for the antenna elements. The antenna was constructed with a very high level of precision. For example, a tolerance of 0.1 mm was used for matching the face of the array. While keeping one of the frequencies constant, the array has the ability to vary the other frequency in small steps (e.g., 30 MHz), from 8.0 to 12.4 GHz.

A 76-element DBF receiver system was reported in [45]. This system was developed by General Electric for Army Missile Command. The system specifications are given in Table 11.9. The stringent image rejection requirements led to the choice of triple down-conversion bandpass sampling. Further, the receivers used digital synchronous phase detectors rather than the more conventional analog detectors in order to produce high-quality I and Q signals. That is, the I and Q signals were both balanced and highly linear. This choice was made with the objective of keeping to minimum the deleterious effects of the residual image components and third-order intermodulation product on the extensive signal processing that takes place in a DBF system.

General Electric developed still another digital beam steering antenna [46]. The detailed specifications for this system are given in Table 11.10.

Chujo et al. [47, 48] carried out a design study on the use of a DBF antenna for mobile satellite communications. They discussed the configurations that the DBF antennas would take if they were used for both receiving and transmitting signals in a digital communications system. A design for an experimental system was presented, consisting of a two-dimensional array with elements arranged in a 2×4 grid pattern. The interelement spacing was a half-wavelength and the antenna polarization was right-hand circular. The system operated at a frequency of 1.54 GHz and used CW signals for both transmit and receive. The calibration of the whole system, except for antenna array, was carried out in the digital domain.

Table 11.9
DBF Receiver Module Characteristics for The System Reported in [45]

Operating freq.	5.25 to 5.85	GHz
Noise figure	6.0 to 7.5	dB
Signal bandwidth	400	KHz
ADC resolution	12	bit
Image rejection	50	dB
Dynamic range	72	dB
Intermodulation products	-50	dB

An eight-element transmit/receive array operating at 1.85 to 1.97 GHz is currently under development at the Communications Research Laboratory, McMaster University, Canada. This DBF system, together with two additional transceivers that act as mobile handsets, will be used to implement a complete two-way digital wireless communications system. Beamforming will be carried out on both transmit and receive. The transceiver will support time-domain duplexing with the option of frequency duplexing as a future upgrade. The receivers are designed for a 4-dB system noise figure, high dynamic range, and 50 dB of gain control. Baseband quadrature outputs with a bandwidth of up to 5 MHz are provided, as well as a second IF output for direct sampling. The maximum transmit power is 25 dBm per element. Data acquisition and processing are handled by a VME/VSB-based system, which consists of an ADC board and a DSP board hosted by a Sparc II class workstation. The acquisition board, an ICS-150, provides four ADC channels, each with a maximum sampling rate of 20 MHz at a 12-bit resolution. Likewise, four 20-MHz D/A channels are available for arbitrary waveform (modulation) generation. Data are buffered at full speed by a large onboard first-in first-out (FIFO) buffer. The acquisition board supports direct-memory-access (DMA) transfer via both VME and VSB buses. A high-speed analog multiplexer will be used to take these signals sampled at the antenna's eight elements and direct them to the signal processor. A high-performance CSIP SC-3 array processor is being used to carry out the DSP functions. This is an i860-based board providing up to 100 MFLOPs. The array processor is programmable in C or FORTRAN and has an extensive vector library. High-speed data transfer in and out of the DSP board is accomplished via both the VME and VSB buses. Digital beamforming and digital receiver functions are handled by the DSP board.

Table 11.10
Parameters for DBF System Reported in [46].

Array Parameters	
Frequency band	5.2 to 5.7 GHz
Array type	32-element linear array
Array element type	8 dipole column +2 dummies
Broadside azimuth HPBW	4.2° (35-dB sidelobe taper)
Elevation HPBW	7.9° (uniform taper)
Azimuth scan capability	±55°
Receiver Parameters	
Receiver type	triple-conversion, frequency-agile
Image rejection	50 dB
Frequency tuning	26 frequencies across band
Short-term stability	10^{-9}
Instantaneous signal bandwidth	500 kHz
Noise figure	4.5 dB
Dynamic range	42 dB
ADCs	10-bit I and Q
Sampling rate	0.5 MHz
ADC operating point	RMS noise to quantizing bit = 2
Phase mismatch	2° rms (max)
dc offset	1/4 LSB
I and Q delay mismatch	20 ns (max)
Calibration method	self-calibrating bilateral loop including synchronous detectors

References

[1] E. O. Brigham, *The Fast Fourier Transform and Its Applications*, Prentice Hall, Englewood Cliffs, NJ, 1988.

[2] J. Mitola, "Software radios," *IEEE Communications Magazine*, May 1995.

[3] P. H. Ladbrooke, *MMIC Design: GaAs FETs and HEMTs*, Artech House, London, 1989.

[4] R. W. Bierig, "Broadband MMICs for system applications," *Microwave J.*, vol. 31, pp. 251–270, May 1988.

[5] M. W. Green et al., "GaAs MMIC yield evaluation," in *17th European Microwave Conf.*, Rome, Italy, 1987.

[6] P. Rigby, J. R. Suffolk and R.S. Pengelly, "Broadband monolithic low noise feedback amplifiers," in *1983 IEEE MTT-S Digest*, 1983.

[7] Y. Ayasli et al., "A monolithic 1-13 GHz traveling wave amplifier," *IEEE Trans. MTT*, vol. 30, July 1982.

[8] J. Eisenberg et al., "A new planar double-double balanced MMIC mixer structure," in *10th Annual IEEE 1991 Microwave and Millimeter-Wave Monolithic Circuits Symposium*, Boston, pp. 69–72, 1991.

[9] P. D. Chow et al., "Ninety-four GHz InAlAs/InGaAs/InP HEMT low noise down converter," in *SPIE-MMIC for Sensors, Radar, and Communications Systems*, Orlando, pp. 48–54, 1991.

[10] H. Inamori et al., "A 2 GHz down converter IC fabricated by an advanced Si bipolar process," *IEEE trans. Consumer Electronics*, vol. 36, pp. 707–711, Aug. 1990.

[11] P. Wallace et al., "MMIC converter cuts of DFS receivers," *Microwave & RF*, vol. 29, 1990.

[12] G. Dietz et al., "A 10-14 GHz quenchable MMIC oscillator," in *10th Annual IEEE 1991 Microwave and Millimeter-Wave Monolithic Circuits Symposium*, Boston, pp. 23–26, 1991.

[13] S. Moghe et al., "High performance GaAs C-band and Ku-band MMIC oscillator," in *IEEE MTT-S Digest*, Las Vegas, pp. 911–914, 1987.

[14] R. Martin and F. Ali, "A Ku-band oscillator subsystem using a broadband GaAs MMIC push-pull amplifier/doubler," *IEEE Microwave and Guided Wave Letters*, vol. 1, pp. 348–350, 1991.

[15] T. Ohira et al., "MMIC 14 GHz VCO and Miller frequency divider for low noise local oscillators," *IEEE Trans. on MTT*, vol. 35, pp. 657–661, July 1987.

[16] S. B. Moghe and T.J. Holden, "High performance GaAs MMIC oscillators," *IEEE Trans. on MTT*, vol. 35, pp. 1283–1287, Dec. 1987.

[17] S. G. Houng et al., "60-70 dB isolation 2-19 GHz MMIC switches," in *11th Annual IEEE GaAs IC Symposium*, San Diego, pp. 173–176, 1989.

[18] J.V. Bellantoni et al., "Monolithic GaAs p-i-n diode switch circuits for high power millimeter-wave applications,," in *IEEE Microwave and Millimeter-Wave Monolithic Circuits Symposium,*, Long Beach, pp. 2162–2165, 1989.

[19] B. Khabbaz et al., "GaAs dc-20 GHz SPDT absorptive switch," in *Proc. Eighth Biennial University/Government/Industry Microelectronics Symposium,*, Westborough, U.S., pp. 165–167, 1989.

[20] A. Ezzeddine et al., "High isolation dc to 18 GHz packaged MMIC SPDT switch," in *18th European Microwave Conf.*, Stockholm, pp. 1028–1033, 1988.

[21] L. Thomas et al., "GaAs MMIC broadband SPDT PIN switch," *Elec. Lett.*, vol. 22, pp. 1183–1185, 1986.

[22] F. Ali et al., "A cell-based wideband monolithic upconverter," in *18th EuMC Digest*, pp. 753–758, 1988.

[23] H. Kato et al., "A 30 GHz MMIC receiver for satellite transponders," *IEEE Trans on MTT*, vol. 38, pp. 896–903, July 1990.

[24] G. Beach et al., "An S-band image rejection receiver front-end incorporating GaAs MMICs," in *15th EuMC Digest*, pp. 1019–1024, 1985.

[25] S. B. Sarjit et al., "An L-band monolithic receiver for military applications," in *18th EuMC Digest*, pp. 463–468, 1988.

[26] J. M. Dieudonne at al., "Advanced MMIC components for Ka-band communications systems. A survey," in *1995 IEEE Microwave Systems Conf.*, Orlando, pp. 11–14, 1995.

[27] G. Codispoti, M. Lisi, and V. Santachiara, "X-band SAR active design for small satellite applications," in *IEEE AP-S Symposium Digest*, pp. 666–669, 1995.

[28] C. Kermarrec et al., "The first GaAs fully integrated microwave receiver for DBS applications at 12 GHz," in *14th EuMC Digest*, pp. 749–754, 1984.

[29] J. Berenz, M. LaCon, and M. Luong, "Single chip Ka-band transceiver," in *1991 IEEE MTT-S Digest*, pp. 517–520, 1991.

[30] M. Moulin et al., "A/D converter characterization for high performance signal application," in *Int. Conf. Analogue to Digital and Digital to Analogue Conversion*, 1991.

[31] F.G. Weiss, "A 1 Gs/s 8-bit GaAs ADC with on-chip current source," in *Proc. GaAs IC Symposium*, p. 209, 1986.

[32] T. Dcourant et al., "3 GHZ, 150mw, 4-bit GaAs analog to digital converter," in *Proc. GaAs IC Symposium*, pp. 301–304, 1986.

[33] R. E. Leonard Jr., "Data converters: getting to know dynamic specs," *Electronic Design*, Nov. 8, 1990.

[34] B. P. Del Signore, "Monolithic 20-b delta-sigma A/D converter implemented in standard 3-um CMOS technology," *IEEE J-SC*, vol. 8, Dec. 1990.

[35] J. Valburg, "8-b 650 MHZ folding A/D converter with analog error correction," *IEEE J-SC*, vol. 10, Dec. 1992.

[36] H. Kimura, "10-b 300 MHz interpolated-parellel ADC implemented in double-polysilicon self-aligned bipolar technology," *IEEE J-SC*, vol. 11, Apr. 1993.

[37] A. N. Karanicolas, "15-b 1-Msample/s digital self-calilbrated pipeline ADC," *IEEE J-SC*, vol. 11, Dec. 1993.

[38] C. Schiller, "4-GHz 8-bit ADC system using sample and filter sampling technique," *IEEE J-SC*, vol. 9, Dec. 1991.

[39] F. Goodenough, "ADCs become application specific," *Electronic Design*, pp. 42–50, Apr. 1993.

[40] R. A. Keihl et al., *High Speed Heterostructure Device*, Academic Press Inc., New York, 1994.

[41] N. Sclater, *Gallium Arsenide IC Technology: Principles and Practice*, TAB Books, Blue Ridge Summit, PA, 1988.

[42] M. Rocchi, *High-Speed Digital IC Technologies*, Artech House, Boston, 1990.

[43] N. Kanopoulos, *Gallium Arsenide Digital Integrated Circuits: A System Perspective*, Prentice Hall, Englewood Cliffs, NJ, 1989.

[44] S. Haykin and V. Kezys, "Multi-parameter adaptive radar system," in *30th Midwest Symposium of Circuits & Systems*, pp. 737–739, 1988.

[45] J. F. Rose, "Digital beamforming receiver technology," in *IEEE AP-S Int. Symposium Digest Antennas Propagat.*, pp. 380–383, 1990.

[46] L. Eber, "Digital beam steering antenna," Technical Report RADC-TR-88-83, General Electric Company, 1988.

[47] W. Chujo and K. YasU.K.awa, "Design study of digital beam forming antenna applicable to mobile satellite communications," in *IEEE AP-S Symposium Digest*, pp. 400–403, 1990.

[48] Y. Ohtaki, et al., "Implementation of a DBF antenna for mobile satellite communications utilizing multi-digital signal processors," in *Proc. 1992 Int. Symposium on Antenna and Propagat.*, Japan, 1992.

Chapter 12

Digital Signal Processing Technology

As mentioned, DBF offers many advantages over its analog counterpart. However, its implementation requires a great deal of raw DSP power. That is, high speed digital processors are needed for operations such as analog-to-digital conversion, addition, multiplication. Therefore, the question that needs to be asked and answered: does DSP technology advance to the point to allow for the cost-effective implementation of DBF in wireless communications systems? To answer this question, we need to first assess the DSP technology in a number of aspects. DSP systems have progressed remarkably in the past decade, especially in the last few years. Advances in circuit technology, architectures and algorithms have led to rapid advances in the capacity of DSP chips, which in turn has led to major improvement in the capabilities of communications systems. The advances are so rapid that they are described in terms of weeks and months. Thus, the assessment provided here is by no means complete and thorough. We are only able to focus our attention on a limited number of issues, which will, we hope, reflect and exemplify the progress of DSP technology.

Based on the rough assessment, we will then try our best to evaluate the feasibility of DSP technology to DBF in different wireless communications systems.

12.1 DSP PROCESSOR ARCHITECTURES

Although in principle all digital signal processing algorithms can be implemented by programming a general purpose digital computer, in many cases it is found that this is not an effective solution.The architectures of general purpose computers are found to be less than optimal when it comes to real time implementation of DSP algorithms. Therefore, the architectures of DSP processors have been developed to have distinct features that distinguish them from general computing systems. One of the major differences between computers and signal processors is the unbalanced distribution of processing operations provided by the latter. This in response to the fact that most signal processing algorithms require a very large number of repeated arithmetic operations of a relatively simple nature, combined with a small number of input/output operations. Furthermore, if real-time operation are desirable, even a modest signal processing need would require a computational rate found only in very large computers, and yet on the other hand, only a small fraction of the sophisticated operating functions that such systems offer would be used. Due to these factors, considerable effort has been invested in the design and implementation of various digital signal processors. Considering the different design methodologies and application environments, DSP processors are usually classified into two main categories: general-purpose processors and special-purpose processors. A generalized structure of a general purpose DSP processor is shown in Figure 12.1. Despite of the wide variety of algorithms used in signal processing, DSP processors usually share a number of common architectural features. This is true for both the general purpose and special purpose DSP processors. The main features of DSP architectures usually consist of fast arithmetic units, multiple functional units, parallel computation, pipelined functional units, localized data communication buses, and high bandwidth buses, etc. In what follows, we will present some of the most common features of DSP processors.

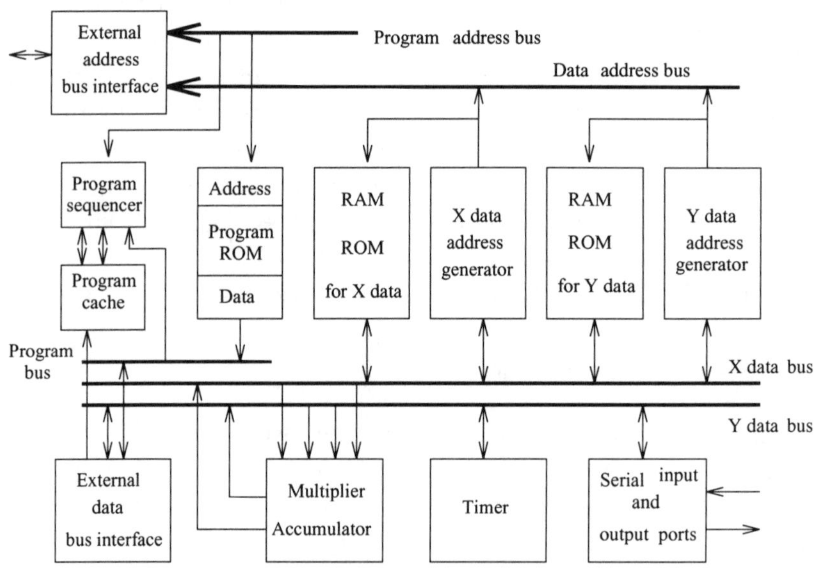

Figure 12.1 Architecture of general purpose digital signal processor.

12.1.1 Specialized Arithmetic Hardware

The most fundamental characteristic of DSP processors is their incorporation and integration of high performance multipliers and adders. Many processors are capable of performing multiply-accumulate operation in a single instruction cycle. Typical multiply-accumulate times can range from 10ns to 100ns.

DSP processors can be categorized by the methods used for arithmetic storage and manipulation: i.e. fixed-point or floating-point. Although fixed-point systems are simpler to design, they suffer from inherent difficulties when it comes to scaling. As the number of adds and multiplies increases, the calculation result will tend to exceed the limitations of the fixed-point scheme. Thus, the results must be scaled to prevent the overflow. In contrast, floating-point systems avoid these limitations because the number is represented by a mantissa and an exponent. However, the disadvantage of floating-point systems is the cost and complexity to support the mantissa-exponent representation.

12.1.2 Multiple Buses

There are many ways to enhance the system throughput by increasing the rate of data access. A common approach is to use multiple memories and corresponding buses for parallel memory access. In general, DSP problems are often easily partitioned into two or more distinct address spaces. For example, when implementing a digital filter, the coefficients, the input and output delay variables, and the program are all distinct entities. By providing three separate address spaces, all three data types can be accessed in parallel. The number and type of multiple buses used in DSP processors varies from device to device. For instance, early processors incorporated a basic Harvard architecture [1], where instructions and data are stored separately in distinct address spaces. Most recently, the architectures of processors have been extended by either providing more than two address spaces, providing a program cache, or a combination of both. For instance, the Motorola DSP96002 provides three address spaces that can be accessed concurrently [2].

Several DSP processors have dual internal data memories and access to an external memory. The ADSP2100 chip from Analog Devices Inc. has a 16-word program cache which allows access to two external data words in a single cycle. The Texas Instrument TMS320C25 chip has a repeat facility [1], which means that a single instruction can be repeated a specific number of times with one single fetch from the external program memory. Moreover, Texas Instrument's TMS320C30 chip is configured with a 64-word program cache and includes a direct memory access (DMA) arrangement which transfers blocks of data from an external memory into the ALU.

12.1.3 Parallelism and Pipelining

The need for parallel processing in DSP systems is becoming increasingly evident. Non-parallel or sequential processors are quickly approaching the upper limit of their computational potential. This potential performance envelope is dictated by the speed of light, which restricts the maximum signal transmission speed in silicon to $3 \mathrm{x} 10^7$ m/s. Thus, a chip with a 3 cm diameter requires of the order of 10^{-9} s. to propagate a signal, thereby restricting the speed of such a chip to at most 1000 MFLOPS. Although other technologies such as gallium arsenide (GaAs) provide a shorter signal propagation time than silicon, they represent only a minor delay to the inevitable saturation of the speed performance of chips. Thus, sequential processors are quickly approaching their upper bound, thereby making parallel computing the wave of the future [3]. Today, many parallel processors incorporate a control-driven or von Neumann form of execution. In this approach, two or more conventional uniprocessors are connected together and concurrently execute traditional fetch-and-execute instructions. In general, there are three types of parallel architectures: single instruction stream with multiple data stream (SIMD), multiple instruction stream with single data stream (MISD), and multiple instruction stream with multiple data stream (MIMD). The SIMD and MIMD have been proved to be the most important control-driven parallel architectures, while MISD has received much less attention and has been challenged as being impractical [4].

The performance of DSP processors is considerably enhanced by considering pipelining with parallel processing. The techniques used by DSP processors can be divided into three categories: interlocking, time-stationary coding, and data-stationary coding. Each is characterized in terms of how the programmer pipelines the parallel execution of instructions [5, 6].

1. When using the interlocking approach, the programmer does not consider the details of the pipelining. That is, the programmer writes code without knowing whether or not the executions of the previous instructions has been completed. If a conflict occurs, such as an instruction requiring results of a previous instruction, the control hardware interlocks the execution of the instruction.
2. In the time-stationary coding approach, the programmer is given more explicit control over the stages of the pipeline. This is accomplished by having each instruction specify all operations that occur simultaneously in a single instruction cycle. When a conflict in access appears, such as a register being used as the destination of an operand fetch and as a source for a multiply-accumulate operation, the source value is taken as that resulting from the earlier instruction making demands on that register.

3. The data-stationary coding approach incorporates instructions that specify all operations to be performed on a single set of operands from memory. Instead of specifying what happens to the hardware at a particular time, in this approach instructions specify what happens to the data throughout its processing cycle. Thus, the results of each instruction are often not immediately available for use by subsequent instructions.

For instance, Motorola's DSP96002 processor uses the time-stationary coding model of pipelined parallel instruction execution [2]. Under this model, the programmer is given explicit control over instructions that occur concurrently in a single instruction cycle and the associated pipeline stages. This processor can achieve a peak performance of 60 MFLOPS with a 40 MHz clock frequency.

12.2 GENERAL AND SPECIAL DSP PROCESSORS

Generally, signal processing is performed using basic mathematical tools in order to get the information of interest into a desired form or representation. The algorithms may appear to be different , but manipulations of the actual signals tend to rely on a relatively small set of fundamental signal processing operations, such as convolution, correlation or difference equation calculations for filtering operation, DFT/FFT coefficient calculation, vector or matrix arithmetic operations, etc. The actual problem at hand simply specifies the way in which these operations are combined to accomplish the required processing and to realize the minimum performance requirements. It is for this reason that general purpose DSP processors have emerged as a major solution to most DSP problems. This is true today as it has been in the past. Table 12.1 are given some of the main features of the most recent commercial DSP processors.

In many applications, only a limited number of highly specialized digital signal processing functions need be implemented. Furthermore, the functions must be performed in real time, and the approach adopted for carrying out the processing must be cost effective and competitive with alternative analog implementations. Sometimes it is found that general purpose DSP processors cannot meet these specialized requirements, while at the same time being cost effective and running in real time. In these instances, special purpose processors arise in a variety of fields, such as telecommunications, radar and sonar processing, to mention only a few. In actual fact, special purpose DSP processors are being developed at a rapid pace as a result in great advances in application specific integrated circuit (ASIC) technology. In general, special purpose DSP processors have lot of features in common with most general purpose processors. The techniques discussed earlier for

Table 12.1
Performance characteristics of commercial DSPs.

Part	*Maker*	*Speed and clock*	*On-chip memory*	*Data width*	*Dedicated I/O*
ADSP21062 SHARC	Analog Device	40 MFLOPS 120 MIPS 40 MHz	32 K words RAM	32 b float	Fast I/O port 240 Mb/s Six link ports 40 Mb/s each Two 40 MHz bidirectional serial ports
TMS320C32	TI	40 MFLOPS 20 MIPS 40 MHz	512 words RAM	32 b float	Two 10 MHz serial ports
TMS320C40	TI	50 MFLOPS 25 MIPS 50 MHz	2 K words RAM	32 b fixed	Six link ports (20 Mb/s)
TMS320C80	TI	theoretical peak 2 GOPS	12 K words	32 b	Crossbar I/O 400 Mb/s
DSP96002	Motorola	60 MFLOPS 20 MIPS 40 MHz	2 K words program 2×512 words data RAM	32 b float	None specific (two external bus ports, used for data and program as well as I/O)
DSP32C	AT&T	25 MFLOPS 12.5 MIPS 50 MHz	1.5 K words RAM	32 b	Serial port, general-purpose buses

achieving high throughput and speed are also employed in the architectures of special purpose DSP processors. Special purpose DSP processors may be programmed in a very narrow sense such as changing the coefficients used in the computation of the algorithm, but they cannot be programmed to solve a completely new problem. In addition, special purpose DSP processors have many architectural features that differ from those of a general purpose DSP. For instance, bus structures may not be, i.e. they may consist of many multiple and non-uniform buses; memories may have multiple ports and may not conform to standard sizes.

To date, a number of special purpose DSPs have been developed, based on the residue number system (RNS) [7, 8] as well as on systolic array technology [9]. The main functions performed by these DSPs include FFTs, inverse FFTs, FIR filtering, convolution, correlation, and vector multiplication [10, 11, 12].

Complex multiplication, which is one of the most fundamental operations, can be carried out using either a residue number system (RNS) arithmetic multiplier or a systolic array multiplier. The multipliers can be realized using an ASIC design. The residue number system, which is 1700-years-old, has been attracting a great deal of attention recently. Digital systems structured into residue arithmetic units may play an important role in ultra-speed, dedicated, real-time systems that support pure parallel processing of integer-valued data. It is a "carry-free" system that performs addition, subtraction, and multiplication as concurrent (parallel) operations, side-stepping one of the principal arithmetic delays — that of managing information that must be carried forward.

The microelectronics revolution has brought low-cost, high-performance RAM and ROM to replace the expensive and slow core that was once used for RNS based systems. Modern semiconductor memory can be programmed to replace traditional arithmetic algorithms by means of a simple table-lookup. Therefore, RNS arithmetic can be performed as a table-lookup mapping, provided the process does not exceed the address space of available memory. Today, RNSs are being built in MOS, TTL, and ECL, and other researchers are exploring novel electro-optical technologies for implementing the RNS [13], but optical implementation is, at best, a wave of the future. Today the semiconductor technology largely decides what advances that are possible in communications.

The RNS is now finding its way into many application areas involving digital filters and mathematical transforms. For example, a two-dimensional, RNS matched filter capable of 20M operations per second was built and tested at Lockheed [14]. A high-speed FFT based on RNS was developed at Martin-Marietta [15]. A group at Mitre has developed error-correction machines using RNS [16]. An RNS comb filter has been reported in [17], while RNS systems implemented using VLSI hardware have been discussed in the open literature [10].

The RNS is potentially very fast even though the advantages of the fixed radix system do not carry over. Algebraic comparison is difficult, as is overflow and sign detection, and division is awkward. An apparent asset of the RNS is its exactness, but it turns out to be a shortcoming. In a weighted-number system, imprecision is induced during truncation or rounding operations—simple operations in a weighted-number system, that are needed to manage potential register overflows. However, because each digit in the RNS is of equal significance, none of the digits can be deleted as can those of a weighted number system. Instead, inefficient RNS division operations must be used for scaling (rounding/truncating) and dynamic-range scaling is required in most RNS applications which involve multiplication. The product of two RNS integers belonging to Z_M is defined in Z_M^2. This geometric increase in dynamic range can rapidly fill any practical dynamic range limit.

The RNS multiplier requires less hardware compared to direct conventional arithmetic. However, it has several drawbacks.

1. To use RNS arithmetic, all inputs must be converted into an RNS representation. The overhead associated with this conversion is approximately equivalent to a conventional multiply operation. Similarly, conversion of final outputs is required to get these quantities back into a conventional binary representation. The associated overhead of which is of the order of three conventional operations.
2. Operations in RNS arithmetic are integer operations. Therefore, cascading of multiplies results in a rapid growth of the output dynamic range.
3. The gain in efficiency using the RNS arithmetic requires that many operations be executed in the RNS system for each associated input and output conversion. Thus, the RNS arithmetic is advantageous only when a hardware intensive implementation is needed for an algorithm that requires many noncascaded multiply operations before an output is obtained.

The systolic multiplier is a distributed-pipelining operation in the direct binary format. Bits in a word are parallelly distributed to separate multipliers and the multiplication is carried out in a pipeline mode. It is interesting in several aspects. First, they are large circuits, where the effect of circuit design as well as the properties of the interconnects affect the circuit performance. Second, they can be implemented by regular structures. Therefore, automatic synthesis techniques can be used. In addition, the interconnect delay can be related directly to the circuit size and shape [18].

Two's-complement systolic array multipliers [19] are very well suited for use in the design of other systolic array structures utilized in DSP applications such as DBF, where multiplication of complex data is required. Because there is no format conversion required, they require less overhead. The multiplication must be implemented using successive additions and the timing between operations are cumbersome. Although this scheme may present the highest permissible clock rate, its implementation becomes increasingly unwieldy as word widths increase. In addition, developing and testing such a time skewed data system can become a formidable task.

2.3 IMPLEMENTATION OF DSP

2.3.1 Semiconductors

ince the introduction of integrated circuits in the early 1960s, and before 1980s, ilicon bipolar technology, particularly emitter-coupled logic (ECL) has been the ominant technology used in the areas of fast computing. Although ECL technol- gy can provide high-speed devices, ECL circuits are characterized by high power issipation and a low level of integration. Moreover, since the desirable character- stics of ECL circuits are directly related to the level at which they consume power, CL cannot benefit from device miniaturization [20, 21] However, the driving capa- ilities of ECL can be improved by using active pull-down circuit technique [20, 22]. he cost that must be sustained is a significant increase in the number of compo- ents.

During the 1980s, complementary metal-oxide-semiconductor (CMOS) emerged s the dominant technology for integrated circuit applications. This trend has con- inued into the 1990s. The characteristics features of CMOS circuits are low cost, ow power dissipation and high level of integration. Simple scaling laws have been sed for reducing FET dimensions [23]. The weakness of CMOS is its relatively ow driving capability. In the 1990s, a new technology called BiCMOS emerged. t combines bipolar and CMOS transistors in a single integrated circuit. By com- ining the strengths of both bipolar and CMOS, BiCMOS is able to achieve ICs vith speed-power-density performance previously unattainable with either bipolar r CMOS technologies individually. Due to the increased complexity of the fabri- ation process, BiCMOS circuits suffer the disadvantages of higher costs and longer abrication time. BiCMOS technology has been used effectively for improving the riving capability, especially at 5 V power supply. However, the advantage of BiC- MOS is being evaded with the introduction of supply voltages that are considerably ower than 5 V [24]. Various schemes have been proposed to cope with this prob- em [25, 26].Simplification of the circuit structures is one of the answers, but much emains to be done in this area. Recently, GaAs technology is now increasingly eing incorporated into high speed digital and analog system design. GaAs tech- ology has been around for more than 25 years, but due to certain difficulties in its nanufacturing process, the development of GaAs has lagged behind that of the sili- on. GaAs devices have many advantages over silicon devices, and this is especially bvious when it comes to high speed applications. GaAs devices are now considered y many system designers as next generation's high speed system components. We vill elaborate on GaAs technology later on.

12.3.2 ASIC Implementation

In recent years, a great deal of attention has been drawn to ASIC (application specific integrated circuit) technology. Emerging as the mainstream of VLSI design, ASIC technology is revolutionizing the design, manufacturing, and marketing practices of electronics-related industries. The essence of this revolution is that in pre-ASIC era, system engineers based their designs on components consisting of microprocessors and standard integrated circuits, a process of building system from chips; whereas in ASIC design, the designer now imports and integrates the system into chips. The advantages of using ASICs include:

1. Normally, the per gate price of an ASIC circuit is less than the equivalent gate price of standard ICs using either *small scale integration* (SSI) or *medium size integration* (MSI). Thus, ASIC parts can reduce semiconductor component costs and overall system costs.
2. Smaller system size is a significant benefit. Compared to using SSI/MSI ASICs can replace 100 or more parts with a single package and operate with less power. This also leads to reduced cooling costs.
3. The ASIC approach can lead to a better systems performance. Because of the application specific orientation, many system overhead functions associated with conventional microprocessors can be eliminated. Thus, the possibility of achieving very high performance is greatly enhanced. Besides, the interconnection delay on circuit boards and backboards can be substantially reduced using ASIC chips.
4. The ASIC approach can offer higher systems reliability. Mechanical connections, such as wire bonds and cables, are a major source of reliability problems. ASIC circuits can reduce mechanical connections by a ratio of 10:1, or more.

Although the full-custom design provides a circuit with optimized performance and density, the semi-custom design has a considerable advantage. The reason is that the semi-custom design can reduce design costs, minimize development times and increase the probability of success. Traditionally, ASIC designs are implemented using two kinds of semi-custom approaches: standard cells and mask programmable gate arrays. Standards cells are basic logic components designed as a customized cells, where all the processing levels are unique. Circuit designers can pick standard cells from the standard cell libraries and pack them together in rows or columns much like variable-height or variable-width building blocks, and then interconnect them to create a high level semi-custom component. On the other hand, the *mask programmable gate array* (MPGA) consists of a standard array of many gate circuits diffused into a chip. Circuit designers provide the manufacturer with a pattern for the interconnecting metallization which will convert the basic gates into functional custom circuits. However, the two main drawbacks of standard cell and MPGA approaches are that:

1. It takes several months to make a prototype after a layout is sent to the factory;
2. It is expensive to make small quantities of chips.

Recently, there has been increasing interest in implementing ASIC design using field-programmable gate arrays (FPGA). The main advantages of FPGA include[27]:

1. It takes only minutes to make a prototype;
2. It is cheap to produce small quantities of chips;
3. It offers low inventory costs and low probability of making mistakes.

In general, the architecture of a FPGA is similar to that of a MPGA, consisting of an array of logic blocks, where the interconnections can be programmable to realize different designs. The major difference between FPGAs and MPGAs is that a MPGA is programmed using integrated circuit fabrication to form metal interconnections while a FPGA is programmed through electrically programmable switches much like the traditional programmable logic device (PLD). However, FPGAs are different from PLDs in the following aspects:

1. The PLD routing architectures are very simple, but use highly inefficient crossbar-like structures in which every output is directly connected to every input through one switch. The FPGA routing architectures provide a more efficient MPGA-like routing where each connection typically passes through several switches.
2. In PLD, logic is implemented using predominately two-level AND-OR logic with wide input AND gates. In FPGA, logic is implemented using multiple levels of lower fan-in gates, which is often more compact than two-level implementations.

Thus, FPGAs can achieve much higher levels of integration than PLDs due to their more complex routing architectures and logic implementations. There are three types of programmable switch technologies used in FPGAs: SRAM, antifuse and EPROM [27]. By choosing the appropriate granularity and functionality of the logic block, and by designing the routing architecture to achieve a high degree of routability while minimizing the number of switches, both the density and performance of FPGA can be optimized [28]. In Tables 12.2 and 12.3, we provide commercially derived assessments of trends in CMOS and ECL gate array technologies. These were obtained from information released LSI Logic and Motorola.

Table 12.2
Commercial CMOS gate array technology from LSI Logic.

	1986	1988	1995	2000
Process technology (μm)	2	1.5	0.35	0.1
Max. no. of usable gates	8,000	50,000	500,000	3,000,000
Gate delay (ns)	1.4	0.57	0.13	0.05
Power (μW/gate/MHz)	20	15	3.6	2

Table 12.3
Commercial ECL gate array technology from Motorola.

	1987	1991	1995	2000
Process technology (μm)	1.5	1	0.7	0.3
Max. no. of usable gates	10,000	70,000	300,000	900,000
Gate delay (ps)	100	40	18	6
Power/gate (mW)	2	1.2	0.6	0.2

12.4 MAJOR FUTURE TRENDS IN DSP

12.4.1 Performance

DSP processor performance will continue to increase geometrically. This means that its performance growth will parallel the historical growth that has occurred in the capacities of general purpose computers. For example, throughout the past decades, Intel has consistently delivered processors with a 4-fold increase in capabilities every three years. Processors based on Hewlett-Packard's precision-architecture reduced-instruction-set-computer (PA-RISC) architecture has averaged performance improvement of 65% per year over the last ten years [29]. By 1996, the microprocessor performance will be nearly 1000 MFLOPS [29]. According to Lewis [30], the performance of processors will surpass the 800-SPECmark near the year 2000, that is, the performance of processors will undergo a tenfold increase in less than a decade. (Note: the 800-SPECmark is a benchmark established by the SPEC consortium of vendors that more or less measures Unix performance under simulated conditions.) Moreover, multiprocessing and parallel processing can offer a further increase in the performance of DSP processors. For instance, the Exemplar Scaleable Parallel Processor, introduced in March 1994 by Convex Computer Corp., which employed as many as 128 HP's CMOS-based PA-RISC processors, with up to 32 Gbits of globally shared physical memory, can attain up to 25000 MFLOPS performance [31].

In the near future, an enormous increase in DSP processor performance may be attributed to:

1. Wide ranging applications of multiprocessing, parallel processing and pipelining in DSP computing;
2. Employment of high speed SRAM cache between the DSP processor and main memory in order to build a well-balanced system [29].

12.4.2 Signal Processing Techniques

The introduction of neural networks is one of most notable recent advances that has occurred in the area of signal processing techniques [32]. Neural networks are composed of parallel, connected processing elements called neurons, which have a large number of inputs but only one output. A weight is stored in each of the inputs. Individual inputs can be thought of as components of a vector, which we may call the input vector. Similarly, the weights can be thought of as a weight vector. Each input is multiplied by its associated weight, and the products are added. This operation is equivalent to taking the inner product of the input vector and the stored weight vector. The result is then made nonlinear by mapping onto a sigmoid function. If the inner product of the input and the weight vector exceeds a specified threshold, the neuron's output moves to a high state; otherwise it remains in the low state. The region around the threshold has a positive slope and is also known as the sigmoid's gain. Many neurons are clustered together and interact to form a network.

The main advantage of a neural network is that it can learn from experience. The other advantages are its parallel computing ability and its capability of handling nonlinear computations.

Interest in using neural networks for signal processing has grown rapidly in recent years, bringing about the first IEEE Workshop in Neural Networks for signal processing in 1991. The most notable features that distinguish neural networks from conventional signal processing techniques are (1) the nonlinear characteristics of artificial neurons, and (2) the learning algorithms that are used to derive the weight vectors. A typical neural network can calculate dot products, the same function that is used in different areas of signal processing, such as discrete Fourier transforms(DFTs) and finite and infinite impulse response filters (FIRs and IIRs). In addition, the neural network can perform certain functions that cannot be performed by linear signal processing systems.

Many DSP algorithms can proceed in a straightforward manner on a neural network combined with a tapped delay line. Adding a tapped delay line to a neural network and eliminating its nonlinear transfer function produce a formal equivalent of a FIR filter.

When using a neural network in a signal processing application, designers need to take advantage of the large body of signal processing knowledge that already exists. For instance, if a low-pass filter function is required, it will be more efficient to use the well-known equations for a FIR filter than to try to train a neural network with a tapped delay line to behave like a low-pass filter.

12.4.3 GaAs Technology

GaAs devices have a history of more than 25 years, however, it is not until recently that they have gained the strong interest of both IC manufacturing and the system's designer communities [33, 34]. As DSP technology is currently developing quickly in the areas high speed and low power consumption, GaAs devices, with their intrinsic characteristics, have great advantages over the now widely used bipolar, CMOS, or BiCMOS technologies.

Transistors in GaAs VLSI switch very swiftly; logic gates based on the GaAs devices impose much less delay on signals than do their silicon counterparts. For example, the shortest silicon CMOS gate delay nowadays is about 150ps versus 70ps for GaAs, both based on 0.5um technology. If we compare the GaAs chips, which are currently available, with silicon chips of comparable functionality, we find that the former are usually a higher performance version of the latter. In other words, the silicon in these chips has been replaced by GaAs in order to take advantage of GaAs's six-fold increase in electron mobility. Still to come in the future are truly novel chips, which will break free of the silicon cage. These will incorporate devices like optical emitters or microwave amplifiers that can only be built using GaAs components.

Another attracting feature of GaAs devices is their low power consumption at high speed. CMOS devices consume very little power at low speed because their static power consumption is very low due to complementary nature of their on-chip structures. But when it comes to high speed, one finds that their dynamic power consumption will increase rapidly and soon become the major component of the device's power consumption. On the other hand, in the case of GaAs devices, the power consumption is almost the same regardless of the operating speed. Of course, it should be noted that GaAs devices only consume less power than CMOS devices at very high frequencies. At lower frequencies, the reverse is usually the case. This is especially true in a system consisting of many ICs, because at any particular time, a fair number of the ICs are idle.

Today's GaAs technology still needs much improvement, both in processes used in the manufacturing of these devices, as well as those used in their design. But one idea seems to be gaining a stronger foothold amongst designers of high speed digital systems, is that GaAs will eventually develop to be the main technology for high speed digital systems. This also holds true for the future of DSP technology.

12.5 SPACE APPLICATION REQUIREMENTS

This section reviews the two major requirements arise from the special space application: radiation hardness and power requirements.

12.5.1 Radiation Hardness Requirements in Space Applications

Satellites and other space systems operate in a high-radiation environment, resulting from trapped particle associated with planetary magnetic fields (e.g. the Van Allen belts which affect earth-orbiting satellites), deep space cosmic rays or high-energy protons from solar flares. This makes radiation hardness one of the major considerations when selecting electronic devices for mobile satellite communication systems [35, 36].

In a low earth orbit, an integrated circuit may be exposed to a few kilorads of radiation over its useful lifetime, whereas at orbits in the middle of the Van Allen belts, exposure levels may increase to several hundred kilorads or more. The radiation level to which a satellite is exposed varies according to its orbital inclination and altitude. A substantial difference also occurs during different periods of solar flare activity. Generally speaking, integrated circuits need to withstand a total dosage of at least 20 kilorads when deployed in a communication satellite. Hardness levels of 100 kilorad or above are preferred.

For integrated circuits to work in such high levels of radiation, radiation hardened technology is required in addition to the conventional method of using shielding material such as aluminum. Radiation hardened integrated circuits are fabricated using specialized processes and designs that increase their tolerance to ionizing radiation levels by several orders of magnitude.

The major radiation threats to ICs are total ionizing dosage and single event phenomena. Ionizing radiation creates bulk-oxide and interface-trap charge that reduces transistor gain and shifts the IC's operating properties, such as the threshold voltage, etc. The total accumulated dose will cause a device to fail if (1) the transistor threshold voltage shifts far enough to cause a circuit malfunction, (2) the device fails to operate at the required frequency, and/or (3) electrical isolation between devices is lost. Single-event phenomena (SEP) occur when a cosmic ray of very high-energy particle impinges on a device. The particle generates a dense track of electron-hole pairs as it passes through the semiconductor, with the free carriers being collected at junctions. The net effect is that circuit is perturbed and may lose data. Increasingly, SEP are becoming of concern to the satellite community, because of the decreasing line widths being used for fabricating ICs, which in turn leads to a corresponding reduction in the cross-sectional areas of transistors, as well as a reduction in the charge stored on the circuit nodes.

Radiation Hardened Technology for ICs

The mainstream semiconductor technologies that are being used nowadays fall into three categories: Bipolar (including ECL etc.), MOS (mainly CMOS), and GaAs. Some of the radiation hardening procedures that are being used by the industry are common to all of these technologies. An example of such a technique is the avoidance of dynamic circuits. On the other hand, other techniques must be tailored to the particular technology under consideration.

Until recently, most of the research into radiation hardening has been devoted to the CMOS. There are a number of reasons for the focus on CMOS. Historically, bipolar circuits have been very tolerant to total ionization dosage. But with the decrease in critical dimension, bipolar circuits become highly susceptible to SEP. Although bipolar technology offers speed advantages over CMOS technology , its relatively high power consumption makes it less desirable then CMOS technology for most space applications. GaAs technology is now considered to be the technology of choice for high speed, high performance, terrestrial-based computer and digital systems. Compared to both bipolar and CMOS technologies, GaAs technology offers low power consumption and high speed performance. GaAs devices are intrinsically more radiation resistant than silicon devices. Researchers have shown that GaAs ICs suffer only minor permanent damage from exposure to high total doses of radiation [37]. However, just as the GaAs technology itself needs a lot of improvement, so does the technology for radiation hardening of GaAs ICs. The most common GaAs circuits structures that are being currently used are either dynamic or very sensitive to SEP in high-radiation environments [38].

CMOS ICs are generally the least sensitive to SEP due to the presence of active devices which restore the original voltage level of a node following a voltage transient induced by a heavy-ion strike. Combined with their low power requirements and high densities, CMOS ICs are often the technology of choice for space applications. In the case of CMOS ICs, the conventional techniques are: (1) thinning gate oxide layers, (2) annealing the gate oxide at a lower temperature, (3) increasing resistance of channel stopper, (3) adopting silicon-on-sapphire (SOS) or silicon-on-insulator(SOI) manufacturing processes [39].

Recently, there has been increased interest in adopting commercial IC technology for space applications [39, 40, 41]. There are many reasons for doing this. First, in some space missions, e.g. satellites in a low earth orbit , the spacecraft is exposed to less radiation during its lifetime. As an example, Space Station Freedom may require integrated circuits with hardening requirements ranging from a few to 20 krads(si) depending on platform location [40]. In these applications, shielding and careful screening of hardening technologies (to take advantage of annealing in the space environment) enables one to use some commercial products. Although most commercial ICs have a radiation hardness of less than 10 krad which is insufficient for a space environment, it is possible by employing certain radiation hardening techniques to reduce commercial ICs, which are capable of operating in space. Secondly, and also the most importantly, the merger of commercial and space/military integrated circuit technologies will offer a number of advantages, including lower cost, reduced development time and procurement time, extensive software support, higher density and performance. Traditionally, radiation hardened ICs are manufactured using separate processes with different technologies and different quality control measures. The market for radiation hardened ICs is minuscule compared to the total commercial market. This — and the time and effort required to produce specialized radiation-hardened ICs — causes these components to be very expensive; therefore, they frequently lag behind the state-of-the-art in capability. It has been estimated that radiation hard CMOS technology development lags about 3 years behind that of the mainstream CMOS technology in general [40, 39].

The Future of Radiation Hardened Technology

The future of radiation hardened technology will depend largely on the development of IC manufacturing technology, as most of the radiation hardened technologies derive from the unique characteristics of the different IC technologies.

Just as GaAs ICs are now becoming recognized as a major choice for high performance terrestrial-based computer and digital systems, we can see radiation hardened GaAs ICs becoming the main choice for high performance space application. In the future, it is expected that improvements to GaAs radiation hardening technology will overcome this technology's susceptibility to SEP.

As radiation hardening technology progresses, we increasingly will see radiation hardened ICs derived from their commercially available ancestors. This will be especially popular with sophisticated, long-development-time ICs such as microprocessors [41].

12.5.2 Power dissipation

Designers are currently able to squeeze more memory and logic onto single chips. However, a major concern has re-emerged, namely, that of power dissipation. This is especially true when it comes to space applications. Processors with 4 to 7 million transistors, using 0.5μm CMOS and operating at clock speeds of several hundred MHz, will consume as much as 35 W even with power supply of 3.3 V. The multiple on-chip parallel operations can drive up the power dissipation to 40 W. IC manufacturers envisage that by the end of this decade processors will have near-zero power dissipation with a power supply voltage of under 1 V.

Because power dissipation is proportional to the square of the supply voltage, the supply voltage must be cut dramatically if there is to be a major reduction in power consumption. For instance, if the supply voltage goes down from 3 V to 1 V, the power dissipation will be reduced by a factor of about 10.

GaAs ICs are able to compete effectively against BiCMOS and CMOS based devices, not only in terms of speed but also integration and overall power dissipation. Due to small logic swings of GaAs and the lower intrinsic capacitance, a GaAs direct-coupled FET logic (DCFL) gate has a lower delay-power product than an equivalent CMOS gate. The DCFL's logic structure is similar to that of NMOS and dissipates static power. However, unlike CMOS, DCFL power dissipation is independent of logic state or operating frequency. Therefore, at high frequencies, GaAs DCFL will dissipate less power than an equivalent CMOS circuit with at the same switching frequency. Besides, GaAs DCFL does not suffer from a reduction in performance at lower voltage levels. Power dissipation can be cut 50% without a corresponding reduction in the technology's performance [42].

According to Vitesse Semiconductor, the speed-power product of GaAs is much better than that of CMOS. Low-power GaAs chips with around 500,000 devices can be fabricated and these chips can be operated at speeds well beyond those of CMOS devices. The cost premium of GaAs chips will eventually become a non-issue because other benefits of GaAs, such as simple chip architectures that are easier to implement and fewer process steps in manufacturing, may make the use of GaAs more cost-effective [42].

12.6 DSP TECHNOLOGY AND DBF

From the preceding sections, we have seen that a single commercially available DSP processor for general purposes offers up to 60 MFLOPS and 120 MIPS. The performance of a microprocessor nearly doubles every year. By 1996, a microprocessor may offer nearly 1000 MFLOPS. A special purposed DSP processor based on ASIC technology can deliver even higher performance. In what follows, we will determine whether it is feasible to implement real-time DBF in a communications system using today's DSP technology.

As we have learned from the early chapters, there are two functions in a DBF system that require high computational power: computation of weights and beamforming. Computation of weights involves the use of an adaptive algorithm to update the beamforming weights. Therefore, the computational complexity is determined by the complexity of the algorithm. For example, the LMS algorithm requires $2N$ complex multiples or $8N$ real multiplies per update per beam, where N is the number of antenna elements. The RLS algorithm requires $4N^2 + 4N + 2$ complex multiples or $16N^2+16N+8$ real multiplies per update per beam per channel. If the update rate is r_u updates per second, the total number of real multiplies required in the RLS case is

$$N_c = r_u \left(16N^2 + 16N + 8\right) \quad \text{per beam per channel} \tag{12.1}$$

Let us consider a simple mobile case where there are $N = 16$ antenna elements. The weights are to be updated every ms ($r_u = 1000$ updates/s), which is equivalent to an update for each wavelength (1 GHz) that a mobile travels at a speed of 100 km/hr. Such a high update rate may be unnecessary in practice, but we use it for the "worst" case estimation. If there are eight co-channel users (i.e., eight beams), the total number of required real multiplies per channel is

$$N_c = 8 \times 1000 \left(16 \times 16^2 + 16 \times 16 + 8\right) = 3.488 \times 10^7/s \tag{12.2}$$

Since a real multiply is approximately equivalent to a floating point operation, 35 MFLOPS is required in this case, which can be easily handled by a commercial DSP processor.

Beamforming involves the multiplication of the sampled signals by the weights. The required complex multiples per beam per channel is the same as the number of antenna elements at the sampling rate. In terms of real multiplies, the total number is given by

$$N_b = 4 \times r_s N \quad \text{per beam per channel} \tag{12.3}$$

where r_s denotes the sampling rate. Let us consider the same case as above and assume that a sampling rate of 3 MHz is used for a 1.5 MHz channel. the total number of real multiplies is

$$N_b = 4 \times 3000000 \times 16 = 1.92 \times 10^8/s \quad \text{per beam per channel} \tag{12.4}$$

or 192 MFLOPS per beam per channel. This number seems large at first. However, one should realize that the beamforming operation is actually the operation of the inner product of two vectors. Therefore, an ASIC can be designed to deliver the required computational performance.

From the above examples, we can see that there should be no major obstacles in the implementation of DBF in land mobile communications systems, voice and low data rate indoor communications systems, and mobile data communications systems. In satellite communications, however, one may anticipate implementation challenges because of the following two reasons:

1. The number of antenna elements is relatively large, thereby increasing the required number of computation;
2. The microprocessor technology for space applications is normally three years behind that for surface applications due to the strict requirements for radiation hardness and power dissipation.

In high data rate communications, one may face similar challenges since the required sampling rate must be relatively high to accommodate the wide bandwidth of the channels.

References

[1] B. Sikstrom et al., "A high speed 2-D discrete cosine transform chip," *Integration, the VLSI Journal*, vol. , pp. 159–169, May 1987.

[2] *DSP96002 IEEE Floating-Point Dual-Port Processor User's Manual*, Motorola Inc., Austin, Texas, 1989.

[3] A.L. DeCegama, *Parallel Processing Architectures and VLSI Hardware*, Prentice-Hall, Englewood Cliffs, NJ, 1989.

[4] K. Hwang and F.A. Briggs, *Computer Architecture and Parallel Processing*, McGraw-Hill, New York, NY, 1984.

[5] E.A. Lee, "Programmable DSP architectures: Part I," *IEEE ASSP Magazine*, vol. , pp. 4–19, October 1988.

[6] E.A. Lee, "Programmable DSP architectures: Part II," *IEEE ASSP Magazine*, vol. , pp. 4–14, January 1989.

[7] D.K. Banerji, "On the use of residue arithmetic for computation," *IEEE Trans. Computers*, vol. C-23, pp. 1315–1317, Dec. 1974.

[8] N.S. Szabo and R.I. Tanaka, *Residue Arithmetic and Its Applications to Computer Technology*, McGraw-Hill, New York, 1967.

[9] H.T. Kung, "Why systolic architectures?," *IEEE Computer*, vol. 1, pp. 37–46, June 1982.

[10] F.J. Taylor, "A VLSI residue arithmetic multiplier," *IEEE Trans. Computers*, vol. C-31, pp. 540–546, June 1982.

[11] D.K. Banerji et al., "A high-speed division method in residue arithmetic," *in IEEE Proc. Fifth Symp. Computer Arithmetic*, pp. 158–164, 1981.

[12] H.T. Kung, "Special-purpose devices for signal and image processing: An opportunity in VLSI," *in Proc. SPIE Symp.*, pp. 76–84, 1980.

[13] A. Huang, "The implementation of a residue arithmetic unit via optical and other physical devices," *in Proc. Int'l Optical Computing Conf.*, pp. 14–18, 1975.

[14] C.H. Huang and F.J. Taylor, "A memory compression scheme for modular arithmetic," *IEEE Trans. Acoustics, Speech and Signal Processing*, vol. ASSP-27, pp. 608–611, Dec. 1979.

[15] W. Smith, "Swift," *in Symp. Very High Speed Computing Technology*, 1980.

[16] D.O. Carhoun et al., "A synthesis algorithm for recursive finite field fir digital filters," Technical report, Mitre technical report, Bedford, Mass., Apr. 1983.

[17] G.A. Jullien, "Residue number scaling and other operations using rom arrays," *IEEE Trans. Computers*, vol. C-27, pp. 325–336, Apr. 1978.

[18] P.J. Song and G.D. Micheli, "Circuit and architecture trade-offs for high-speed multiplication," *IEEE Journal of Solid-State Circuits*, vol. 26, pp. 1184–1198, Sept. 1991.

[19] M.A. Sid-Ahmed, "Two's-complement systolic array multiplier with applications to dsp hardware," *Int.J. Electronics*, vol. 66, pp. 507–518, 1989.

[20] K. Chin et al., "Sub-15 ps large-buffered active-pull-down ECL/NTL circuits," *in*

Proc. 1992 IEEE Custom Integrated Circuits Conf., pp. 7.6.1–7.6.4, 1992.

[21] C.T. Chuang et al., "High-speed low-power ECL circuit with AC coupled selfbiased dynamic current source and active pull down emitter follower stage," *in Proc. 1992 IEEE Custom Integrated Circuits Conf.*, pp. 7.7.1–7.7.4, 1992.

[22] C.T. Chuang et al., "On the leverage of high-f_t transistors for advanced high-speed bipolar circuits," *IEEE J. Solid-State Circuits*, vol. 27, pp. 225–228, February 1992.

[23] R. Dennard et al., "Design of ion-implanted MOSFET's with very small physical dimensions," *IEEE J. Solid-State Circuits*, vol. 9, pp. 256–268, 1974.

[24] L. Wissel and E.L. Gould, "Optimal usage of CMOS within a BiCMOS technology," *IEEE J. Solid-State Circuits*, vol. 27, pp. 300–306, March 1992.

[25] C.L. Chen, "2.5V bipolar/CMOS circuits for 0.25 μm BiCMOS technology," *in 1991 Int. Symp. VLSI Circuits Dig. Tech. Papers*, pp. 121–122, 1991.

[26] M. Hiraki et al., "A 1.5V full-swing BiCMOS logic circuit," *in 1992 Int. Solid-State Circuits Conf. Dig. Tech. Papers*, pp. 48–49, 1992.

[27] J.S. Rose and S. Brown, "Flexibility of interconnection structures for field-programmable gate arrays," *IEEE J. Solid-State Circuits*, vol. 26, pp. 277–282, March 1991.

[28] S. Singh et al., "The effect of logic block architecture on FPGA performance," *IEEE J. Solid-State Circuits*, vol. 27, pp. 281–287, March 1992.

[29] P. Bemis, "Performance in the year 2000: More than clock frequency," *Computer Design*, vol. 34, pp. 82, January 1995.

[30] T.G. Lewis, "Where is computing headed?," *IEEE Computer*, vol. , pp. 59–63, August 1994.

[31] G. Khermouch, "Technology 1995: Large computers," *IEEE Spectrum*, vol. , pp. 48–51, January 1995.

[32] B. H. Juang, S.Y. Kung, and C.A.Kamm, eds., *Proc. IEEE Workshop Neural Networks for Signal Processing, IEEE Press, Piscataway*, 1991.

[33] Ira Deyhimy, "Gallium Arsenide Joins the Giants," *IEEE Spectrum*, vol. , pp. 33–40, Feb 95.

[34] Neil Sclater, *Gallium Arsenide IC Technology: Principles and Practice*, TAB Books Inc., 1988.

[35] R. Koga et al., "Heavy ion induced upsets in semiconductor devices," *IEEE Trans. Nuc. Sci.*, vol. NS-32, pp. 159, Feb 1985.

[36] G. R. Agrawl et al., "A proposed SEU tolerant dynamic random access memory(DRAM) cell," *IEEE Trans. Nuc. Sci.*, vol. 41, pp. 2035, Dec 1994.

[37] M. A. Listvan, "Ionizing radiation hardness of GaAs technology," *IEEE Trans. Nuc. Sci.*, vol. NS-34, pp. 1664, Dec 1987.

[38] D. J. Fouts et al., "Single event upset in Gallium Arsenide dynamic logic," *IEEE Trans. Nuc. Sci.*, vol. 41, pp. 2244, Dec 1994.

[39] M. Yoshioka et al., "A radiation-hardened 32-bit microprocessor based on the commercial CMOS process," *IEEE Trans. Nuc. Sci.*, vol. 41, pp. 2481, Dec 1994.

[40] J. L. Kaschmitter et al., "Operation of commercial R3000 processors in the low earth orbit environment," *IEEE Trans. Nuc. Sci.*, vol. 38, pp. 1415, Dec 1991.

[41] J. R. Kimbrough et al., "Single event effects and performance prediction for space applications of RSIC processors," *IEEE Trans. Nuc. Sci.*, vol. 41, pp. 2706, Dec 1994.

[42] C. Gardner, "Low-power GaAs offers a high-performance alternative to CMOS," *Electronic Design*, vol. 43, pp. 57, January 1995.

Glossary

1-D	one-dimensional
2-D	two-dimensional

A

ABF	adaptive beamforming
ADC	analog-to-digital converter
AMPS	advanced mobile phone service
AOA	angle of arrival
ARDIS	advanced radio data information system
ASIC	application specific integrated circuit
AWGN	additive white Gaussian noise

B

BER	bit error rate
BFSK	binary frequency-shift keying
bps	bits per second
BPSK	binary phase-shift keying
BW	beamwidth

C

CAB	cyclic adaptive beamforming
cdf	cumulative density function
CDM	code-division multiplex
CDMA	code division multiple access
CDPD	cellular digital packet data
CIR	carrier-to-interference ratio
CM	constant modulus
CMA	constant modulus algorithm
CMOS	complementary metal oxide semiconductor
CSMA	carrier sense multiple access
CW	continuous wave

D

DAC	digital-to-analog converter
DAMA	demanded-assigned multiple access
DBPSK	differential binary phase-shift keying
DBF	digital beamforming
DBS	direct broadcast satellite
dc	direct current
DCA	dynamic channel assignment
DCFL	direct-coupled FET logic
DDBM	double-double balanced mixer
DDC	digital down-converter
DECT	digital European cordless telephone
DFT	discrete Fourier transform
DMA	direct memory access
DQPSK	differential quadrature phase-shift keying
DRA	directly radiating array
DRAF	dual-reflector array-fed
DRAM	dynamic random-access memory
DRO	dielectric resonator oscillator
DSB-AM	double sideband amplitude modulation
DSMA	data sense multiple access
DSP	digital signal processing
DSSS	direct-sequence spread spectrum

E

ECL	emitter-coupled logic
EIRP	effective isotropic radiation power
ESPRIT	Estimation of Signal Parameters by Rotational Invariance Technique

F

FBLP	forward-backward linear prediction
FDD	frequency-division duplexing
FDM	frequency-division multiplex
FDMA	frequency-division multiple access
FET	field-effect transistor
FFT	fast Fourier transform
FIFO	first-in first-out
FIR	finite impulse response

FM	frequency modulation
FPGA	field programmable gate array

G

GEO	geo-synchronous orbit
GMSK	Gaussian minimum shift keying
GPS	Gobal Positioning System
GSM	Gobal System for mobile communications
G/T	gain-to-noise-temperature ratio

H

HEMT	high-electron-mobility transfer
HMIC	hybrid microwave integrated circuit
HPA	high power amplifier
HPBW	half-power beamwidth

I

IC	integrated circuit
IF	intermediate frequency
IIR	infinite impulse response
INR	interference-to-noise ratio
ISI	intersymbol interference
ISR	interference-to-signal ratio

K

kbps	kilobits per second

L

LAN	local area networks
LEO	low earth orbit
LLSE	linear least square error
LMS	least mean squares
LNA	low-noise amplifier
LO	local oscillator

M

MCPC	multichannel per carrier
MEO	medium-altitude earth orbit
MESFET	metal semiconductor field-effect transistor

MFLOPS	million floating point operations per second
MIC	microwave integrated circuit
MIMD	multiple-instruction stream with multiple-data stream
MIPS	million instructions per second
MISD	multiple-instruction stream with single-data stream
MLE	maximum likelihood estimation
MLM	maximum likelihood method
MMIC	monolithic microwave integrated circuit
MPGA	mask programmable gate array
MSAT	mobile satellite
MSE	mean-square error
MSI	medium-size integration
MUSIC	multiple signal classification
MVDR	minimum variance distortionless response

N

NCC	network control center

P

PA-RISC	precision-architecture reduced-instruction-set computer
PCM	pulse code modulation
PCS	personal communications services
pdf	probability density function
PEGS	principal eigenvector Gram-Schmidt
PHP	personal handy phone
PLD	programmable logic device
PLL	phase-locked loop
PM	phase modulation
PSK	phase shift keying

Q

QAM	quadrature amplitude modulation
QPSK	quadrature phase-shift keying

R

RAM	random-access memory
RF	radio frequency
RISC	reduced-instruction-set computer
RLS	recursive least squares

rms	root mean square
ROM	read-only memory
RNS	residue number system

S

SAW	surface acoustic wave
SCORE	spectral self-coherence restoral
SCM	sample covariance matrix
SCPC	single-channel-per-carrier
SDMA	space division multiple access
SEP	single-event phenomena
SER	symbol error rate
SHA	sample and hold amplifier
SIMD	single-instruction stream with multiple-data stream
SINR	signal-to-interference-plus-noise ratio
SIR	signal-to-interference ratio
SLC	sidelobe canceler
SMI	sample-matrix inversion
SNR	signal-to-noise ratio
SQNR	signal-to-quantization-noise ratio
SRFF	single-reflector focal-fed
SRHT	single-reflector with hybrid transform
SUPB	single user per beam
SS	satellite-switched
SSI	small-size integration
SSPA	solid-state power amplifier
SWR	standing wave ratio

T

TCM	trellis-coded modulation
TD	threshold detector
TDD	time-division duplexing
TDMA	time-division multiple access
TDM	time-division multiplex
TE	transverse electric
TEM	total equivalent mass
TETRA	Trans-European Trunked Radio
TM	transverse magnetic
TWTA	travling-wave-tube amplifier
T/R	transmitter/receiver

ABOUT THE AUTHORS

John Litva has spent some 30 years in radio engineering. He graduated with a PhD in physics from the University of Western Ontario. For his PhD dissertation he determined the characteristics of large traveling ionospheric disturbances using enhanced radio sources in the sun corona and a ground-based interferometer. From there he has gone on to develop expertise in the following areas: (i) wireless communications (ii) low-angle radar tracking, (iii) microstrip antennas and arrays, (iv) electromagnetic modeling, (v) radio-wave propagation, and (vi) digital beamforming. He has worked on smart antennas over the course of more than 20 years, first for tracking low-angle missiles and most recently for applications in wireless communications. During his low-angle tracking studies, he became an expert in over-sea multipath. More recently, he has become an expert in indoor wireless propagation. While a governments scientist, he established a world class group in the area of low-angle radar tracking. More recently at McMaster University, he has established a world class antenna and microwave laboratory. He is a thrust leader for the Telecommunications Research Institute (TRIO) of Ontario in the area of Antennas and Signal Processing. TRIO is a Provincial Center of Excellence, linking four universities and 28 high-tech companies. He is also the director of the Communications Research Laboratory, which is a leading university-based research institute in the area of telecommunications.

Titus K. Y. Lo received his B.A.Sc. degree from the University of British Columbia, B. C., and M. Eng. and Ph.D. degrees from McMaster University, all in electrical engineering. His areas of research interest include advanced signal processing techniques and their applications in wireless communications and remote sensing. Dr. Lo has been with the Communications Research Laboratory, McMaster University, Ontario, Canada for more than 10 years. He is a senior research engineer at present. He has been involved in numerous research and development projects, including digital beamforming for mobile communications satellites, digital beamforming for spaceborne synthetic aperture radar systems, adaptive beamforming for personal communications services, and intelligent signal processing for wireless communications. He has published more than 50 technical papers and reports on selected topics in antenna array signal processing, propagation, and communications.

Index

The Artech House Mobile Communications Series

John Walker, Series Editor

Advanced Technology for Road Transport: IVHS and ATT, Ian Catling, editor

An Introduction to GSM, Siegmund M. Redl, Matthias K. Weber, Malcolm W. Oliphant

CDMA for Wireless Personal Communications, Ramjee Prasad

Cellular Communications: Worldwide Market Development, Garry A. Garrard

Cellular Digital Packet Data, Muthuthamby Sreetharan, Rajiv Kumar

Cellular Mobile Systems Engineering, Saleh Faruque

Cellular Radio: Analog and Digital Systems, Asha Mehrotra

Cellular Radio: Performance Engineering, Asha Mehrotra

Cellular Radio Systems, D. M. Balston, R. C. V. Macario, editors

Digital Beamforming in Wireless Communications, John Litva, Titus Kwok-Yeung Lo

GSM System Engineering, Asha Mehrotra

Handbook of Land-Mobile Radio System Coverage, Garry C. Hess

Introduction to Wireless Local Loop, William Webb

Introduction to Radio Propagation for Fixed and Mobile Communications, John Doble

Land-Mobile Radio System Engineering, Garry C. Hess

Low Earth Orbital Satellites for Personal Communication Networks, Abbas Jamalipour

Mobile Antenna Systems Handbook, K. Fujimoto, J. R. James

Mobile Communications in the U.S. and Europe: Regulation, Technology, and Markets, Michael Paetsch

Mobile Data Communications Systems, Peter Wong, David Britland

Mobile Information Systems, John Walker, editor

Personal Communications Networks, Alan David Hadden

RF and Microwave Circuit Design for Wireless Communications, Lawrence E. Larson, editor

Smart Highways, Smart Cars, Richard Whelan

Spread Spectrum CDMA Systems for Wireless Communications, Savo G. Glisic, Branka Vucetic

Transport in Europe, Christian Gerondeau

Understanding GPS: Principles and Applications, Elliott D. Kaplan, editor

Vehicle Location and Navigation Systems, Yilin Zhao

Wireless Communications for Intelligent Transportation Systems, Scott D. Elliott, Daniel J. Dailey

Wireless Communications in Developing Countries: Cellular and Satellite Systems, Rachael E. Schwartz

Wireless Data Networking, Nathan J. Muller

Wireless: The Revolution in Personal Telecommunications, Ira Brodsky

For further information on these and other Artech House titles, including previously considered out-of-print books now available through our In-Print-Forever™ (IPF™) program, contact:

Artech House
685 Canton Street
Norwood, MA 02062
781-769-9750
Fax: 781-769-6334
Telex: 951-659
e-mail: artech@artech-house.com

Artech House
Portland House, Stag Place
London SW1E 5XA England
+44 (0) 171-973-8077
Fax: +44 (0) 171-630-0166
Telex: 951-659
e-mail: artech-uk@artech-house.com

Find us on the World Wide Web at: www.artech-house.com